A TEXTBOOK ON WIRELESS COMMUNICATIONS AND NETWORKS

A TEXTBOOK ON WIRELESS COMMUNICATIONS AND NETWORKS

(II-Semester of B.Tech. ECE and M. Tech. for Embedded Systems)
As per JNTU Syllabus

By

K. SRINIVASA REDDY
M. Tech., IAENG,
Associate Professor and Head Department of
Electronics and Communication Engineering,
Chilkur Balaji Institute of Technology
Hyderabad, Andhra Pradesh

LAXMI PUBLICATIONS (P) LTD
An ISO 9001:2015 Company
BENGALURU • CHENNAI • GUWAHATI • HYDERABAD
KOCHI • KOLKATA • LUCKNOW • RANCHI
NEW DELHI

Printed and bound in India *First*
Edition : 2011, 2015; *Edition* : 2025
ISBN : 978-93-80856-58-2

Branches

✆	Bengaluru	080-26 75 69 30
✆	Chennai	044-24 34 47 26
✆	Guwahati	0361-254 36 69
✆	Hyderabad	040-27 55 53 83
✆	Kochi	0484-405 13 03
✆	Kolkata	033-40 04 77 79
✆	Lucknow	0522-430 36 13
✆	Ranchi	0651-224 24 64

Published in India by

UNIVERSITY SCIENCE PRESS
(An Imprint of Laxmi Publications Pvt. Ltd.)

An ISO 9001:2015 Company
113, GOLDEN HOUSE, GURUDWARA ROAD, DARYAGANJ,
NEW DELHI - 110002, INDIA

Corporate Office : B-33, Sector-5, Noida,
Gautambuddha Nagar (U.P.) - 201 301, INDIA
Telephone : +91-120-44 46 900
www.laxmipublications.com
info@laxmipublications.com

CONTENTS

PREFACE

Wireless Communications plays a major role in our lives. Wireless networking products are rapidly emerging and promise to become a major part of the telecommunications infrastructure within the next decade. Developments in Wireless Communication Technology have increased its applications in allied fields of electronics, security, military and defence.

Every care has been taken to eliminate misprints, omissions and errors but it is too much to expect that no inaccuracy/misprint/error has crept in and the author would be grateful to the readers for bringing to his notice any such errors/misprints/omissions they may come across while going through the book.

The author will like to thank the readers (both teachers and students) who sent their valuable suggestions that became the basis for the revision of the book.

The author shall feel satisfied if the book meets the needs of the students for whom it is meant. Suggestions and criticism for improvement of the book are always welcomed.

The author hopes that book will fulfill the need of interested readers and welcome any suggestions towards the improvement of the book.

—**Author**

ACKNOWLEDGEMENT

I would like to thank *Prof.M.Moorthy Muthukrishnan* for his valuable Supervision and support during the preparation of this book. The author extends his thanks to all my *M.Tech* classmants for their valuable suggestions.

I must express my deep appreciation to my wife *Mrs. Geetha Reddy* and my family for their patience, co-operation and understanding which made the tedious task of writing this book much easy to endure.

Special thanks to Dr. Sudhakar Rao P (Dean R and D, ECE, VITS-HYD), Mr. M. Jagadesh Chandra Prasad (professor, BVRIT), Mr. David Soloman Raju.Y (Associate Professor, ECE, HIET-HYD) for constant encouragement.

Great care has been taken to produce this book free from errors, but some may remain.

I will therefore accept the pleasure any information, which will draw the attention to such mistakes and errors.

Criticism about the book shall always be appreciated.

—Author

WIRELESS COMMUNICATIONSAND NETWORKS

UNIT I

WIRELESS COMMUNICATIONS & SYSTEM FUNDAMENTALS: Introduction to wireless communications systems, examples, comparisons & trends. Cellular concepts-frequency reuse, strategies, interference & system capacity, trucking & grade of service, improving coverage &capacity in cellular systems.

UNIT II

MULTIPLE ACCESS TECHNIQUES FOR WIRELESS COMMUNICATION: FDMA, TDMA, SSMA (FHMA/CDMA/Hybrid techniques), SDMA technique (AS applicable to wireless communications). Packet radio access-protocols, CSMA protocols, reservation protocols, capture effect in packet radio, capacity of cellular systems.

UNIT III

WIRELESS NETWORKING: Introduction, differences in wireless & fixed telephone networks, traffic routing in wireless networks–circuit switching, packet switching X.25 protocol.

UNIT IV

Wireless data services—cellular digital packet data (CDPD),advanced radio data information systems, RAM mobile data (RMD). Common channel signaling (CCS), ISDN-Broad band ISDN & ATM-Signaling System no. 7(SS7)-protocols, network services part, user part, signaling traffic, services & performance.

UNIT V

MOBILE IP AND WIRELESS APPLICATION PROTOCOL: Mobile IP Operation of mobile IP, Co-located address, Registration, Tunneling, WAP Architecture, overview, WML scripts, WAP service, WAP session protocol, wireless transaction, Wireless datagram protocol.

UNIT VI

WIRELESS LAN TECHNOLOGY: Infrared LANs, Spread spectrum LANs, Narrow bank microwave LANs, IEEE 802 protocol Architecture, IEEE802 architecture and services, 802.11 medium access control, 802.11 physical layer.

UNIT VII

BLUE TOOTH: Overview, Radio specification, Base band specification, Links manager specification, Logical link control and adaptation protocol. Introduction to WLL Technology.

UNIT VIII

MOBILE DATA NETWORKS: Introduction, Data oriented CDPD Network, GPRS and higher data rates, Short messaging service in GSM, Mobile application protocol.

TEXTBOOKS

1. Wireless Communication and Networking—William Stallings, PHI, 2003.
2. Wireless Communications, Principles, Practice—Theodore, S. Rappaport, PHI, 2nd Edn., 2002.
3. Principles of Wireless Networks—Kaveh Pah Laven and P. Krishna Murthy, Pearson Education, 2002.

REFERENCE

1. Wireless Digital Communications—Kamilo Feher, PHI, 1999.

UNIT 1 WIRELESS COMMUNICATIONS AND SYSTEM FUNDAMENTALS

1.1 INTRODUCTION

The rapid evolution of cellular technology and the more user demand for advanced mobile services leads the industry to develop more efficient network structures. The increasing number of cellular subscribers and the demand for cellular networks communication (3rd and 4th generation) drives to the research of the new methodologies for the design of cellular networks and services.

1.2 EXAMPLES OF WIRELESS COMMUNICATION SYSTEMS

Most people are familiar with a number of mobile radio communication systems used in everyday life. Garage door openers, remote controllers for home entertainment equipment, cordless telephones,hand-held walkie-talkies, pagers (also called paging receivers or "beepers"), and cellular telephones are all examples of mobile radio communication systems. However, the cost, complexity,performance, and types of services offered by each of these mobile systems are vastly different.The term *mobile* has historically been used to classify *any* radio terminal that could be moved during operation. More recently, the term *mobile* is used to describe a radio terminal that is attached to a high speed mobile platform (*e.g.*, a cellular telephone in a fast moving vehicle) whereas the term *portable* describes a radio terminal that can be hand-held and used by someone at walking speed (*e.g.*, a walkie-talkie or cordless telephone inside a home). The term *subscriber* is often used to describe a mobile or portable user because in most mobile communication systems,each user pays a subscription fee to use the system, and each user's communication device is called a *subscriber unit.* In general, the collective group of users in a wireless system are called *users* or *mobiles,* even though many of the users may actually use portable terminals. The mobiles communicate to fixed *base stations* which are connected to a commercial power source and a fixed *backbone network*. Mobile radio transmission systems may be classified as *simplex, half-duplex* or *full-duplex*. In simplex systems, communication is possible in only one direction. Paging systems, in which messages are received but not acknowledged, are simplex systems. Half-duplex radio systems allow two-way communication, but use the same radio channel for both transmission and reception. This means that at any given time, a user can only transmit or receive information. Constraints like "push-to-talk" and "release-to-listen" are fundamental features of half-duplex systems. Full duplex systems, on the other hand, allow simultaneous radio transmission and reception between a subscriber and a base station, by providing two simultaneous but separate channels (Frequency Division Duplex, or FDD) or adjacent time slots on a single radio channel (Time Division Duplex, or TDD) for communication to and from the user.

In FDD, a pair of simplex channels with a fixed and known frequency separation is used to define a specific radio channel in the system. The channel used to convey traffic to the mobile user from a base station is called the *forward channel,* while the channel used to carry traffic from the mobile user to a base station is called the *reverse channel.* In the U.S. AMPS standard, the reverse channel has a frequency which is exactly 45 MHz lower than that of the forward channel. Full duplex mobile radio systems provide many of the capabilities of the standard telephone, with the added convenience of mobility. Full duplex and half-duplex systems use *transceivers* for radio communication.

In Time Division Duplexing (TDD) uses the fact that it is possible to share a single radio channel in time, so that a portion of the time is used to transmit from the base station to the mobile, and the remaining time is used to transmit from the mobile to the base station. If the data transmission rate in the channel is much greater than the end-user's data rate, it is possible to store information bursts and provide the appearance of full duplex operation to a user, even though there are *not* two simultaneous radio transmissions at any instant. TDD is only possible with digital transmission formats and digital modulation, and is very sensitive to timing. It is for this reason that TDD has only recently been used, and only for indoor or small area wireless applications where the physical coverage distances (and thus the radio propagation time delay) are much smaller than the many kilometers used in conventional cellular telephone systems.

Basic Definitions in Cellular System

Base Station	A fixed station in a mobile radio system used for radio communication with mobile stations. Base stations are located at the center or on the edge of a coverage region and consist of radio channels and transmitter and receiver antennas mounted on a tower.
Control Channel	Radio channel used for transmission of call setup, call request, call initiation, and other beacon or control purposes.
Forward Channel	Radio channel used for transmission of information from the base station to the mobile.
Full Duplex System	Communication systems which allow simultaneous two-way communication. Transmission and reception is typically on two different channels (FDD) although new cordless/PCS systems are using TDD.
Half Duplex Systems	Communication systems which allow two-way communication by using the same radio channel for both transmission and reception. At any given time, the user can only either transmit or receive information.
Handoff	The process of transferring a mobile station from one channel or base station to another.
Mobile Station	A station in the cellular radio service intended for use while in motion at unspecified locations. Mobile stations may be hand-held personal units (portables) or installed in vehicles (mobiles).

Mobile switching Center	Switching center which coordinates the routing of calls in a large service area. In a cellular radio system, the MSC connects the cellular base stations and the mobiles to the PSTN. An MSC is also called a Mobile Telephone Switching Office (MTSO).
Page	A brief message which is broadcast over the entire service area, usually in a simulcast fashion by many base stations at the same time.
Reverse Channel	Radio channel used for transmission of information from the mobile to base station.
Roamer	A mobile station which operates in a service area (market) other than that from which service has been subscribed.
Simplex Systems	Communication systems which provide only one-way communication.
Subscriber	A user who pays subscription charges for using a mobile communications system.
Transceiver	A device capable of simultaneously transmitting and receiving radio signals.

1.3 COMPARISON OF COMMON WIRELESS COMMUNICATION SYSTEMS

Comparison of Mobile Communication Systems—Mobile Station

Service	*Coverage Range*	*Required Infra-Structure*	*Complexity*	*Hardware Cost*	*Carrier Frequency*	*Functionality*
TV Remote Control	Low	Low	Low	Low	Infrared	Transmitter
Garage Door Opener	Low	Low	Low	Low	< 100 MHz	Transmitter
Paging System	High	High	Low	Low	< 1 GHz	Receiver
Cordless Phone	Low	Low	Moderate	Low	< 1 GHz	Transceiver
Cellular Phone	High	High	High	Moderate	< 2 GHz	Transceiver

Comparison of Mobile Communication Systems—Base Station

Service	*Coverage Range*	*Required Infra-Structure*	*Complexity*	*Hardware Cost*	*Carrier Frequency*	*Functionality*
TV Remote Control	Low	Low	Low	Low	Infrared	Receiver
Garage Door Opener	Low	Low	Low	Low	< 100 MHz	Receiver
Paging System	High	High	High	High	<1 GHz	Transmitter
Cordless Phone	Low	Low	Low	Moderate	<1 GHz	Transceiver
Cellular Phone	High	High	High	High	<2 GHz	Transceiver

1.4 TRENDS IN CELLULAR RADIO AND PERSONAL COMMUNICATIONS

Since 1989, there has been enormous activity throughout the world to develop personal wireless systems that combine the network intelligence of today's PSTN with modern digital signal processing and RF technology. The concept, called Personal Communications Services (PCS), originated in the United Kingdom when three companies were given spectrum in the 1800 MHz range to develop Personal Communication Networks (PCN) throughout Great Britain.

PCN was seen by the U.K. as a means of improving its international competitiveness in the wireless field while developing new wireless systems and services for citizens. Presently, field trials are being conducted throughout the world to determine the suitability of various modulation,multiple-access, and networking techniques for future 3G PCN and PCS systems.The terms PCN and PCS are often used interchangeably. PCN refers to a wireless networking concept where any user can make or receive calls, no matter where they are, using a light-weight, personalized communicator. PCS refers to new wireless systems that incorporate more network features and are more personalized than existing cellular radio systems, but which do not embody all of the concepts of an ideal PCN.

Indoor wireless networking products are rapidly emerging and promise to become a major part of the telecommunications infrastructure within the next decade. The European Telecommunications Standard Institute (ETSI) is also developing the 20 Mbps HIPERLAN standard for indoor wireless networks. Breakthrough products such as Motorola's 18 GHz Altair WIN (wireless information network) modem, that was not commercialized, and Avaya's (formerly NCR and Lucent) /ORiNOCO waveLAN computer modem have been available as wireless ethernet connections since 1990 and are beginning to penetrate the business world. As we enter the 21st century, products are emerging that allow users to link their phone with their computer within an office environment, as well as in a public setting, such as an airport or train station. A worldwide standard, the Future Public Land Mobile Telephone System (FPLMTS)- renamed International Mobile Telecommunication 2000 (IMT-2000) in mid-1995—has been formulated by the International Telecommunications Union (ITU) which is the standards body for the United Nations, with headquarters in Geneva, Switzerland. The technical group TG 8/1 standards task group is within the ITU's Radiocommunications Sector (ITU-R). ITU-R was formerly known as the "Consultative Committee for International Radiocommunications (CCIR)". TG 8/1 is considering how worldwide wireless networks should evolve and how worldwide frequency coordination might be implemented to allow subscriber units to work anywhere in the world. FPLMTS (now IMT-2000) is a third generation universal, multi-function, globally compatible digital mobile radio system that will integrate paging, cordless, and cellular systems, as well as Low Earth Orbit (LEO) satellites, into one universal mobile system. A total of 230 MHz in frequency bands 1885 to 2025 MHz and 2110 to 2200 MHz was targeted by the ITU's 1992 World Administrative Radio Conference (WARC). In March 1999, ITU-R agreed to additional spectrum allocations that include the frequency bands 806 to 960 MHz, 1710 to 2200 MHz, and 2520 to 2670 MHz. This additional spectrum allocation was approved in May 2000 at the ITU World Radio Conference (WRC-2000). The types of modulation, speech coding, and multiple access schemes to be used in IMT-2000 were also locked down by the ITU Radio General Assembly in mid-2000. The selected radio interfaces for terrestrial wireless service include expansions of today's GSM and IS-95 CDMA, as well as a new timecode CDMA standard proposed by China.Worldwide standards are also required for Low Earth Orbit (LEO) satellite communication systems that were developed in the 1990s but which failed commercially at the turn of the century. Due to the very large areas on earth which are illuminated by satellite transmitters, satellitebased cellular systems will never approach the capacities provided by land-based microcellular systems. However, satellite mobile systems were touted as offering tremendous promise for paging, data collection, and emergency communications, as well as for global roaming. In early 1990, the aerospace industry demonstrated the first successful launch of a small satellite on a rocket from a jet aircraft. This launch technique was pioneered by Orbital Sciences Corp. and was more than an order of magnitude less expensive than conventional ground-based launches and allowed rapid deployment, suggesting that a network of LEOs could be rapidly launched for wireless communications around the globe. While several companies, such as Motorola's Iridium, proposed systems and service concepts for worldwide paging, cellular telephone, and emergency navigation and notification in the early 1990s [IEE91], the capital markets have not supported mobile satellite systems in general. In emerging nations, where existing telephone service is almost nonexistent, fixed cellular telephone systems are being installed at a rapid rate. This is due to the fact that developing nations are finding it is quicker and more affordable to install cellular telephone systems for fixed home use, rather than install wires in neighbourhoods which have not yet received telephone connections to the PSTN.

1.4.1 Paging Systems

Paging systems are communication systems that send brief messages to a subscriber. Depending on the type of service, the message may be either a numeric message, an alphanumeric message, or a voice message. Paging systems are typically used to notify a subscriber of the need to call a particular telephone number or travel to a known location to receive further instructions. In modern paging systems, news headlines, stock quotations, and faxes may be sent. A message is sent to a paging subscriber via the paging system access number (usually a toll-free telephone number) with a telephone keypad or modem. The issued message is called a *page*. The paging system then transmits the page throughout the service area using base stations which broadcast the page on a radio carrier.Paging systems vary widely in their complexity and coverage area. While simple paging systems may cover a limited range of 2 to 5 km, or may even be confined to within individual buildings,wide area paging systems can provide worldwide coverage. Though paging receivers are simple and inexpensive, the transmission system required is quite sophisticated. Wide area paging systems consist of a network of telephone lines, many base station transmitters, and large radio towers that simultaneously broadcast a page from each base station (this is called *simulcasting*). Simulcast transmitters may be located within the same service area or in different cities or countries. Paging systems are designed to provide reliable communication to subscribers wherever they are; whether inside a building, driving on a highway, or flying in an airplane. This necessitates large transmitter powers (on the order of kilowatts) and low data rates (a couple of thousand bits per second) for maximum coverage from each base station. Figure 1.1 shows a diagram of a wide area paging system.

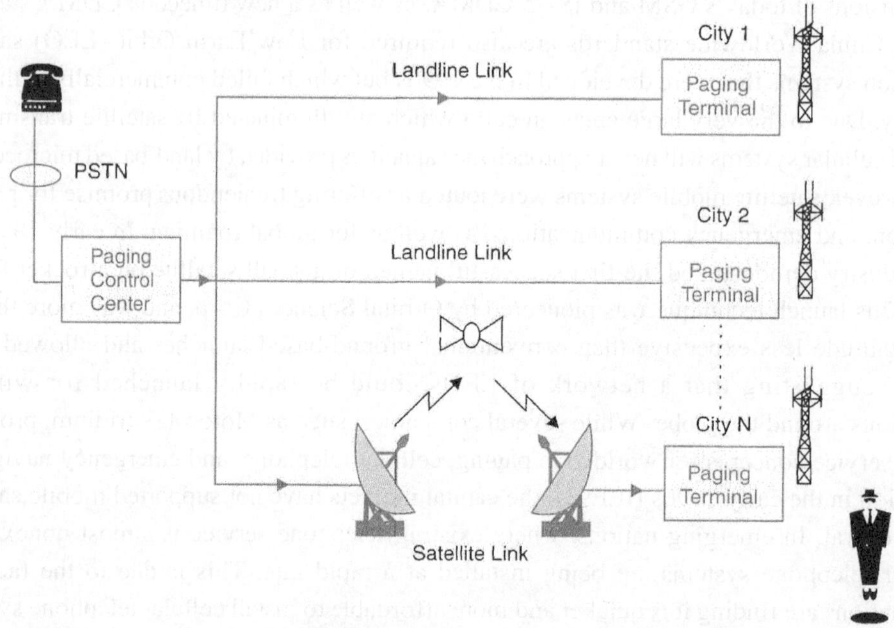

Fig. 1.1. A wide area paging system. The paging control center dispatches pages received from the PSTN throughout several cities at the same time

1.4.2 Cordless Telephone Systems

Cordless telephone systems are full duplex communication systems that use radio to connect a portable handset to a dedicated base station, which is then connected to a dedicated telephone line with a specific telephone number on the Public Switched Telephone Network (PSTN). In first generation cordless telephone systems (manufactured in the 1980s), the portable unit communicates only to the dedicated base unit and only over distances of a few tens of meters. Early cordless telephones operate solely as extension telephones to a transceiver connected to a subscriber line on the PSTN and are primarily for in-home use. Second generation cordless telephones have recently been introduced which allow subscribers to use their handsets at many outdoor locations within urban centers such as London or Hong Kong.Modern cordless telephones are sometimes combined with paging receivers so that a subscriber may first be paged and then respond to the page using the cordless telephone. Cordless telephone systems provide the user with limited range and mobility, as it is usually not possible to maintain a call if the user travels outside the range of the base station. Typical second generation base stations provide coverage ranges up to a few hundred meters. Figure 1.2 illustrates a cordless telephone system.

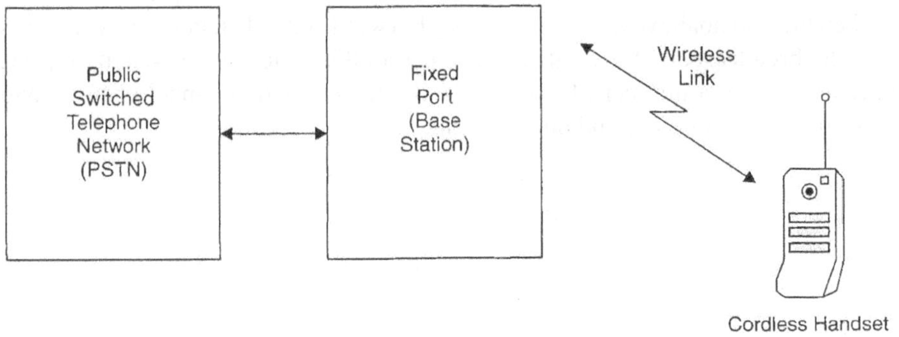

Fig. 1.2. A cordless telephone system

1.4.3 Cellular Telephone Systems

A cellular telephone system provides a wireless connection to the PSTN for any user location within the radio range of the system. Cellular systems accommodate a large number of users over a large geographic area, within a limited frequency spectrum. Cellular radio systems provide high quality service that is often comparable to that of the landline telephone systems. High capacity is achieved by limiting the coverage of each base station transmitter to a small geographic area called a *cell* so that the same radio channels may be reused by another base station located some distance away. A sophisticated switching technique called a *handoff* enables a call to proceed uninterrupted when the user moves from one cell to another. Figure 1.3 shows a basic cellular system which consists of mobile stations, base stations and a Mobile Switching Center (MSC). The mobile switching center is sometimes called a Mobile Telephone Switching Office (MTSO), since it is responsible for connecting all mobiles to the PSTN in a cellular system. Each mobile communicates via radio with one of the base stations and may be handed-off to any number of base stations throughout the duration of a call. The mobile station contains a transceiver, an antenna, and control circuitry, and may be mounted in a vehicle or used as a portable hand-held unit. The base stations consist of several transmitters and receivers which simultaneously handle full duplex communications and generally have towers which support several transmitting and receiving antennas. The base station serves as a bridge between all mobile users in the cell and connects the simultaneous mobile calls via telephone lines or microwave links to the MSC. The MSC coordinates the activities of all of the base stations and connects the entire cellular system to the PSTN. A typical MSC handles 100,000 cellular subscribers and 5,000 simultaneous conversations at a time, and accommodates all billing and system maintenance functions, as well. In large cities, several MSCs are used by a single carrier.

Communication between the base station and the mobiles is defined by a standard Common Air Interface (CAI) that specifies four different channels. The channels used for voice transmission from the base station to mobiles are called Forward Voice Channels (FVC), and the channels used for voice transmission from mobiles to the base station are called Reverse Voice Channels (RVC).The two channels responsible for initiating mobile calls are the Forward Control Channels (FCC) and Reverse Control Channels (RCC). Control channels are often called setup channels because they are only involved in setting up a call and moving it to an unused voice channel. Control channels transmit and receive data messages that carry call initiation and service requests, and are monitored by mobiles when they do not have a call in progress. Forward control channels also serve as beacons which continually broadcast all of the traffic requests for all mobiles in the system. Supervisory and data messages are sent in a number of ways to facilitate automatic channel changes and handoff instructions for the mobiles before and during a call.

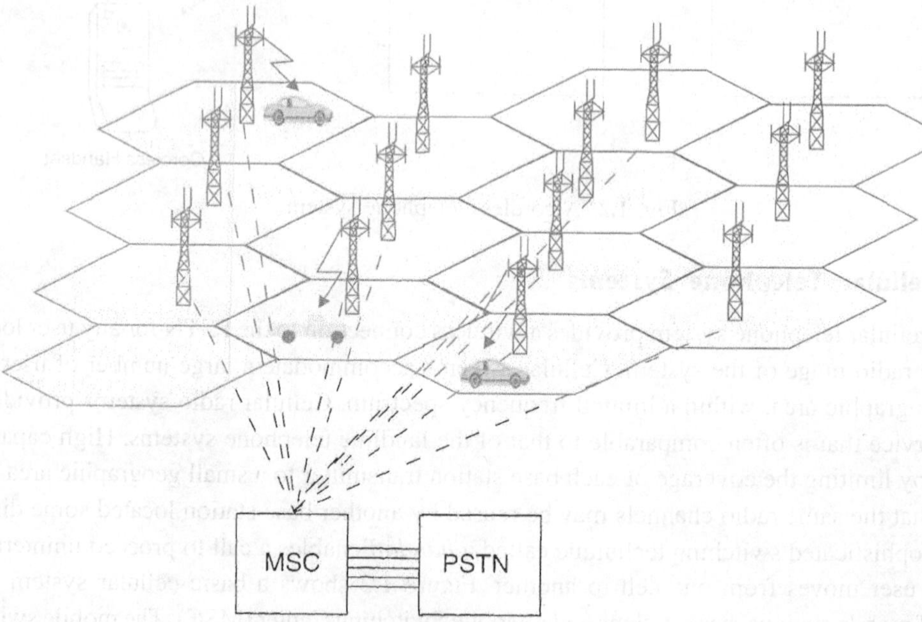

Fig. 1.3. A cellular system. The towers represent base stations which provide radio access between mobile users and the Mobile Switching Center (MSC)

1.5 GENERAL CELLULAR CONCEPTS

In a cellular system the hand-sets carried by the users are called Mobile Stations (MS). The Mobile Stations communicate to the Base Stations (BS) through a pair of frequency channels, one for up-link and another for down-link. All the base stations of a Cellular systems are controlled by a central switching station called Mobile Switching Center (MSC) or Mobile Telephone Switching Office (MTSO). The MSC is responsible for all kinds of network management functions such as channel allocations, Handoffs, billing, power control etc. The MSC is also connected to the Public Subscriber Telephone Network (PSTN) or Public Land Mobile Network (PLMN) so as to allow the MS to talk to a Land Line telephone or vice versa.

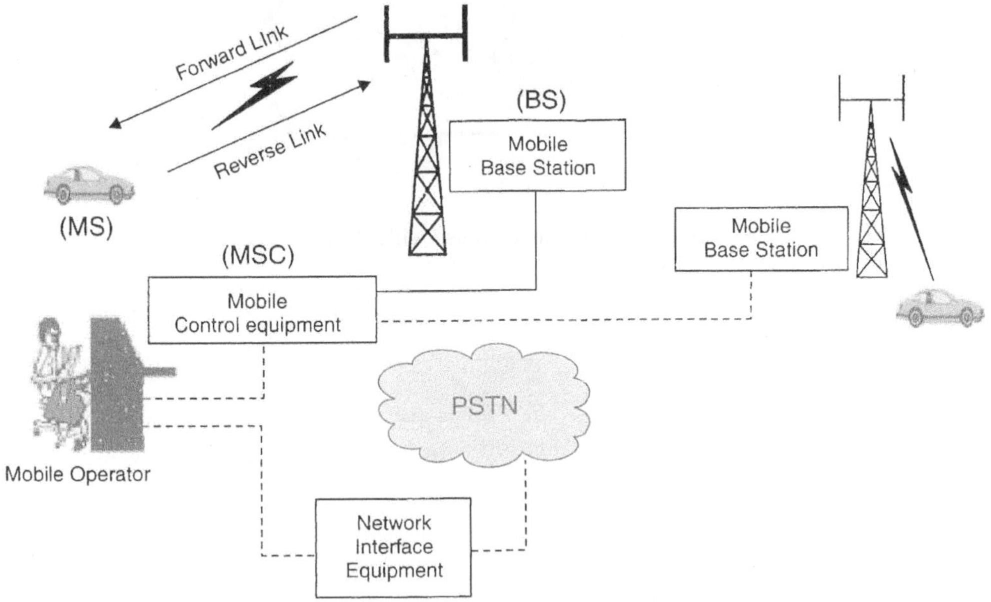

Fig. 1.4. Cellular architecture

Generally two range of frequencies are allocated to Mobile Cellular Systems one for up-link channels and other for down-link channels. This range of frequencies is divided into many channels so as to allow many MS to talk simultaneously. This technique of multiplexing several users over a given spectrum is called Frequency Division Multiple Access (FDMA). Other techniques like Time Division Multiple Access (TDMA) and Code Division Multiple Access (CDMA) are also employed by modern cellular systems. To support a large number of users in a given area a technique called "Frequency Reuse" is used by cellular systems. Under this technique the available channels are divided into smaller groups and a group of channels is allocated to a small region called "cell", typically of Hexagonal shape. In this way the whole range of frequency channels is used over a small group of cells and this group of cells is called "cluster". Typical cluster sizes are 4,7,9,12 and 19.

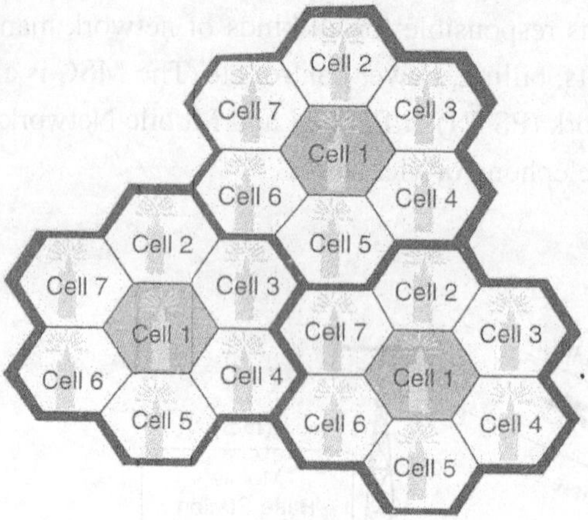

Fig. 1.5. Frequency reuse with cluster

UNIT 2 MULTIPLE ACCESS TECHNIQUES FOR WIRELESS COMMUNICATION

2.1 INTRODUCTION

The radio frequency spectrum, a finite natural resource, has greater demands placed on it every day. In an effort to make the most efficient use of this resource, various technologies have been developed so that multiple, simultaneous users can be supported in a finite amount of spectrum. This concept is called **"multiple accesses."** The four commonly used access methods are Frequency Division Multiple Access (FDMA), Time Division Multiple Access (TDMA), Code Division Multiple Access (CDMA), and SDMA.

Wireless communications technologies, which were virtually unheard of as recently as the late 1970s, are now prevalent throughout today's society and growing in demand. However, limited by the finite frequency spectrum, spectrally efficient technologies have not kept pace with today's high traffic demands. To ensure profit grows parallel with the demand for wireless technologies, manufacturers have had to develop methods of putting more users in the same spectrum space.

The functionality of each access method (FDMA, TDMA, CDMA, SDMA) the advantages and disadvantages of each technology, and various forms of implementation for each technology. FDMA and TDMA are currently being used to support conventional and trunked radio systems, as well as commercial cellular systems. CDMA is being used primarily in cellular systems at this time.

An easy to understand example is given to provide the reader a general overview of how each technology differs. Next, a comparison is made between the technologies followed by a discussion of the primary benefits each technology offers to wireless communication system. Each radio frequency or "channel" must consist of enough space to carry the intended signal to its destination. Therefore, when the terms "spectrum space," "radio frequencies," or "channels" are used, it should be understood that actual physical space is involved, albeit invisible to the human eye. Radio signals vary in size depending on the type of "message" being carried. The size of the signal is referred to as "bandwidth."

Multiple Access Techniques for Wireless Communications

(1) *Random Access Techniques*

- ALOHA Scheme
- Carrier Sense Multiple Access with Collision Avoidance (CSMA/CA) Scheme

(2) *Collision-free Access Techniques*
- Frequency Division Multiple Access (FDMA)
- Time Division Multiple Access (TDMA)
- Code Division Multiple Access (CDMA)
- Spectral Efficiency of FDMA, TDMA and CDMA Systems

There are several techniques how multiple accessing can be achieved. These include the following four basic schemes.

1. Frequency Division Multiple Access (FDMA)
2. Time Division Multiple Access (TDMA)
3. Code Division Multiple Access (CDMA)
4. Space Division Multiple Access (SDMA)

2.1.1 Frequency Division Multiple Access (FDMA)

FDMA is one of the earliest multiple-access techniques for cellular systems when continuous transmission is required for analog services. In this technique the bandwidth is divided into a number of channels and distributed among users with a finite portion of bandwidth for permanent use as illustrated in Fig. 2.1. The channels are assigned only when demanded by the users. Therefore when a channel is not in use it becomes a wasted resource. FDMA channels have narrow bandwidth (30 kHz) and therefore they are usually implemented in narrowband systems. Since the user has his portion of the bandwidth all the time, FDMA does not require synchronization or timing control, which makes it algorithmically simple. Even though no two users use the same frequency band at the same time, guard bands are introduced between frequency bands to minimize adjacent channel interference. Guard bands are unused frequency slots that separate neighbouring channels. This leads to a waste of bandwidth. When continuous transmission is not required, bandwidth goes wasted since it is not being utilized for a portion of the time. In wireless communications, FDMA achieves simultaneous transmission and reception by using Frequency Division Duplexing (FDD). In order for both the transmitter and the receiver to operate at the same time, FDD requires duplexers. The requirement of duplexers in the FDMA system makes it expensive. Frequency division is the original multiple access technique. Currently, most legacy public safety wireless networks use FDMA to improve spectrum efficiency. FDMA is used throughout the commercial wireless industry. Legacy commercial telecommunication networks (analog networks based on Advanced Mobile Phone Service [AMPS] and Total Access Communications System [TACS] standards) are built on a backbone of cellular base stations, using the FDMA technology. However, due to increased spectrum efficiency of CDMA and TDMA systems, very few, if any, new cellular systems are using FDMA.

How it Works

FDMA systems separate a client's large frequency band into several smaller individual bands/ channels. Each channel has the ability to support a user. Guard bands are used to separate channels to prevent interference. They are used to isolate channels from adjacent-channel interference.

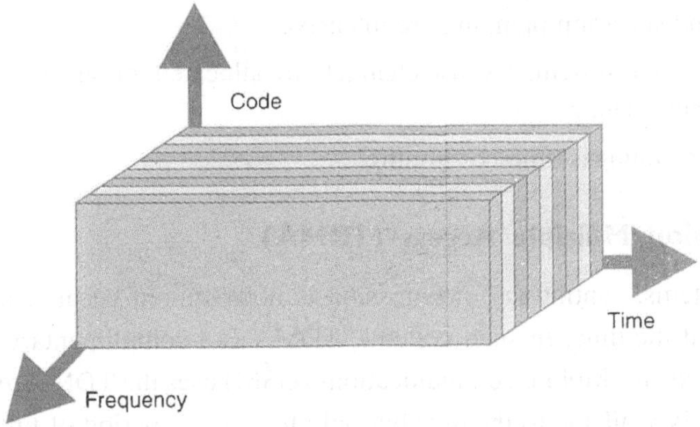

Fig. 2.1 Channel usages by FDMA

FDMA includes the following features:

- In FDMA, one phone circuit is carried by one FDMA channel at a time.
- In FDMA, if an FDMA channel is not in use, it cannot be used by other users to increase the channel capacity. It remains idle.
- FDMA is implemented in Narrowband systems, because, in FDMA, each channel supports only one circuit per carrier so the bandwidth of FDMA channels are relatively narrow (30 KHz).
- Inter symbol Interference is low in Narrowband systems because Narrowband signal has large symbol time as compare to delay spread.
- FDMA mobile systems are less complex as compare to TDMA mobile systems.
- FDMA is a continuous transmission scheme as compare to TDMA because fewer bits are needed for synchronization and framing.
- In FDMA, as unique channels are assigned to each user, so FDMA systems have higher cell site system cost as compared to TDMA system.
- In FDMA, both the transmitter and receiver operates at the same time so FDMA
- Mobile units require duplexers. These also increase the cost of FDMA subscriber units and base station.

Advantages/Disadvantages

Advantages

- Simple to implement, from a hardware standpoint
- Fairly efficient with a small base population and when traffic is constant.

Disadvantages

- Network and spectrum planning are intensive
- In a conventional system, because channels are allocated for one user, idle channels add to spectrum inefficiency
- Frequency planning is time-consuming.

2.1.2 Time Division Multiple Access (TDMA)

In digital systems, continuous transmission is not required because users do not use the allotted bandwidth all the time. In such systems, TDMA is a complimentary access technique to FDMA. Global Systems for Mobile communications (GSM) uses the TDMA technique. In TDMA, the entire bandwidth is available to the user but only for a finite period of time. In most cases the available bandwidth is divided into fewer channels compared to FDMA and the users are allotted time slots during which they have the entire channel bandwidth at their disposal. This is illustrated in Fig. 2.2. TDMA requires careful time synchronization since users share the bandwidth in the frequency domain. Since the number of channels are less, inter channel interference is almost negligible, hence the guard time between the channels is considerably smaller. Guard time is a spacing in time between the TDMA bursts. In cellular communications, when a user moves from one cell to another there is a chance that user could experience a call loss if there are no free time slots available. TDMA uses different time slots for transmission and reception. This type of duplexing is referred to as Time Division Duplexing (TDD). TDD does not require duplexers.

How it Works

Similar to FDMA trunked system, when a user depresses the Push-To-Talk (PTT) switch in a TDMA system, a control channel registers the radio to the closest base station. During registration, the base station assigns the user an available pair of channels, one to transmit and the other to receive. But, unlike an FDMA system registration, a TDMA system registration also assigns an available time-slot within the channel. The user can only send or receive information at that time, regardless of the availability of other time-slots. Information flow is not continuous for any user, but rather is sent and received in bursts. The bursts are re-assembled at the receiving end and appear to provide continuous sound because the process is very fast.

Because TDMA systems also split an allotted portion of the frequency spectrum into smaller slots (channels), they require the same level of frequency planning as FDMA systems. The same careful steps in frequency planning must be taken in both FDMA and TDMA systems.

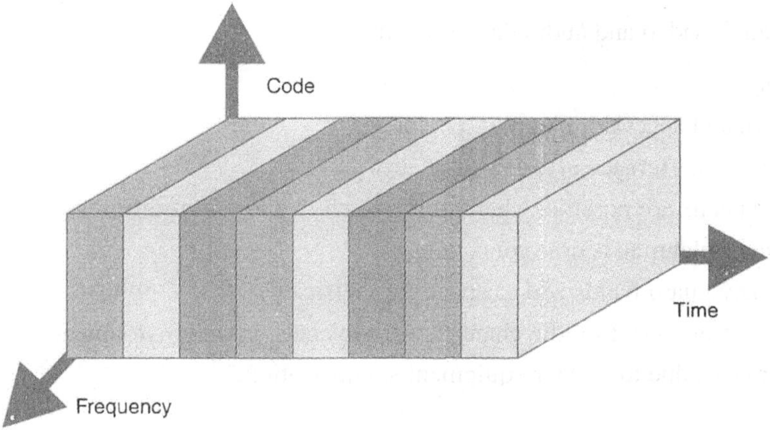

Fig. 2.2. Channel usages by TDMA

TDMA includes the following features:

- In TDMA, by using no overlapping time slots several users shares a single carrier frequency. The number of time slots per frame depends upon several factors, such as modulation techniques and bandwidth.
- In TDMA, data transmission is not continuous and subscriber transmitter can be turned off which result in low battery consumption.
- In TDMA, handoff process is much simpler for a subscriber because of discontinuous transmission.
- In TDMA, duplexers are not required because different timeslots are used for transmission and reception.
- In TDMA, the rate of transmission is very high as compare to FDMA.
- In TDMA, synchronization is required and the time for synchronization or guard time should be minimum.
- TDMA systems require large overheads (framing bits) as compare to FDMA because of discontinuous or burst transmission.
- In TDMA, bandwidth can be utilized on demand of different users as different numbers of time slots are allocated to different users.

Advantages/Disadvantages

Advantages

- Extended battery life and talk time
- More efficient use of spectrum, compared to FDMA
- Will accommodate more users in the same spectrum space than an FDMA system which improves capacity in high traffic areas, such as large metropolitan areas

- Efficient utilization of hierarchical cell structures—pico, micro, and macro cells
- Can handle video and audio data efficiently.

Disadvantages

- Network and spectrum planning are intensive
- Multipath interference affects call quality
- Dropped calls are possible when users switch in and out of different cells
- Frequency planning is time consuming
- Frequency guard bands add to spectrum inefficiency
- Too few users result in idle channels (rural versus urban environment)
- Higher costs due to greater equipment sophistication.

2.1.3 Code Division Multiple Access

In CDMA, all the users occupy the same bandwidth, however they are all assigned separate codes, which differentiates them from each other as shown in Fig. 2.3(*a*). Fig. 2.3(*b*) shows typical Walsh codes used for this purpose. CDMA systems utilize a spread spectrum technique in which a spreading signal, which is uncorrelated to the signal and has a large bandwidth, is used to spread the narrow band message signal. Direct Sequence Spread Spectrum (DS-SS) is most commonly used for CDMA. In DS-SS, the message signal is multiplied by a Pseudo Random Noise Code (PN code), which has noise-like properties. Each user has his own codeword which is orthogonal to the codes of other users. In order to detect the user, the receiver is required to know the codeword used by the transmitter. Unlike TDMA, CDMA does not require time synchronization between the users. A CDMA system experiences a problem called self-jamming which arises when the spreading codes used for different users are not exactly orthogonal. While dispreading, this leads to a significant contribution from other users to the receiver decision statistic. If the power of the multiple users in a CDMA system is unequal, then the user with the strongest signal power will be demodulated at the receiver. The strength of the received signal raises the noise floor for the weaker signals at the demodulators. This reduces the probability that weaker signals will be received. This problem, known as the near-far problem can be taken care of by using power control. This ensures that all the signals within the coverage of the base station arrive with same power at the receiver. CDMA is a spread spectrum technique used to increase spectrum efficiency over current FDMA and TDMA systems. Although spread spectrum's application to cellular telephony is relatively new, it is not a new technology. Spread spectrum has been used in many military applications, such as anti-jamming (because of the spread signal, it is difficult to interfere with or jam), ranging (measuring the distance of the transmission to determine when it will be received), and secure communications (the spread spectrum signal is very hard to detect).

How it Works

With CDMA, unique digital codes (Walsh Codes), rather than separate radio frequencies/ channels, are used to differentiate users. The Walsh codes are shared by the mobile phone and the base station, and are called "pseudo-Random Code Sequences." All users access the entire spectrum allocation all of the time. That is, every user uses the entire block of allocated spectrum space to carry his/her message.

CDMA, being a "spread-spectrum" technology, spreads the information contained in a signal over the entire available bandwidth and not simply through one frequency. Due to the wide bandwidth of a spread-spectrum signal, it is very difficult to cause jamming, difficult to interfere with, and difficult to identify. It appears as nothing more than a slight rise in the "noise floor" or interference level, unlike other technologies where the power of the signal is concentrated in a narrower band making it easier to detect. Therefore CDMA systems provide more privacy than FDMA or TDMA systems. These are great advantages over technologies using a narrower bandwidth.

CDMA channels can handle an unspecified number of users. There is not a fixed number. The capacity of the system depends on the quality of current calls. As more users are added, noise is added to the wideband frequency and therefore decreases the quality of current calls. Each user's transmission power increases the level of the frequency spectrum's "noise floor" and therefore decreases the overall call quality for all users. To help eliminate the "noise floor," CDMA mobile phones and base stations use the minimum amount of power required to Communicate with each other. They use precise power control to decrease users' transmission power. By decreasing a user's transmission power, the mobile phone has added battery life, increased talk time, and smaller batteries.

Because CDMA is a spread spectrum technology, it requires less frequency planning. The full original spectrum is not divided into separate blocks/channels, like it is in FDMA and TDMA systems. Therefore, there is no need to plan for multiple frequency guard bands. Because all users have access to the entire spectrum at all times, frequency planning only needs to consider one frequency/channel. However, the channel requires relatively wide contiguous bandwidth.

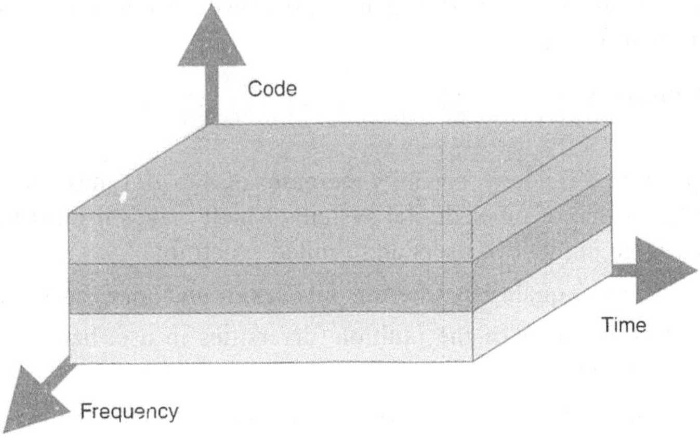

Fig. 2.3(*a*). Channel usage by CDMA

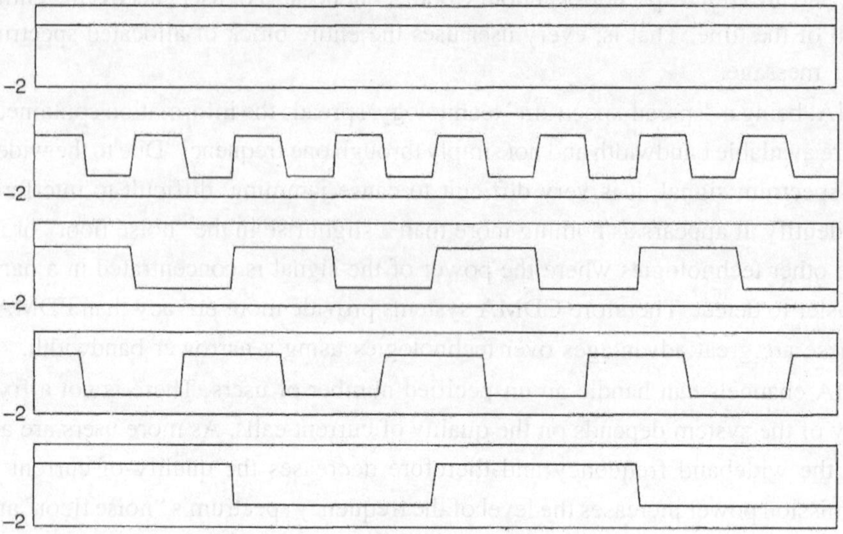

Fig. 2.3(*b*). Walsh codes

CDMA includes the following features:

- In CDMA system many users share the same frequency.
- In CDMA unlike FDMA and TDMA the number of users is not limited. It has a soft capacity. But due to large number of users its performance degrades.
- In CDMA, the spread spectrum or bandwidth is large so multipart fading is reduced.
- In CDMA, the channels data rates are very high.
- CDMA systems provides soft handoff and this handoff is performed by the MSC.
- In CDMA systems, if spreading sequences of different users are not exactly orthogonal, the Self-jamming problem is arises.
- In CDMA systems, if a desired user has less power as compare to undesired users, the near-far problem occurs.

Advantages/Disadvantages

Advantages

- Greatest spectrum efficiency: capacity increases of 8 to 10 times that of an analog system and 4 to 5 times that of other digital systems which makes it most useful in high traffic areas with a large number of users and limited spectrum
- CDMA improves call quality by filtering out background noise, cross-talk, and interference
- "Soft handoffs—Because of the multiple diversities in use, handoffs between cells are undetected by the user
- Simplified frequency planning—all users on a CDMA system use the same radio frequency spectrum.
  - Engineering detailed frequency plans are not necessary.
  - Frequency re-tunes for expansion are eliminated.

- Fewer cells are required for quality coverage

- Random Walsh codes enhance user privacy; a spread-spectrum advantage
- Precise power control increases talk time and battery size for mobile phones.

Disadvantages

- Backwards compatibility techniques are costly
- Currently, base station equipment is expensive
- Difficult to optimize to maximize performance
- Low traffic areas lead to inefficient use of spectrum and equipment resources.

2.1.4 Space Division Multiple Access (SDMA)

SDMA utilizes the spatial separation of the users in order to optimize the use of the frequency spectrum. A primitive form of SDMA is when the same frequency is re-used in different cells in a cellular wireless network. However for limited co-channel interference it is required that the cells be sufficiently separated. This limits the number of cells a region can be divided into and hence limits the frequency re-use factor. A more advanced approach can further increase the capacity of the network. This technique would enable frequency re-use within the cell. It uses a Smart Antenna technique that employs antenna arrays backed by some intelligent signal processing to steer the antenna pattern in the direction of the desired user and places nulls in the direction of the interfering signals. Since these arrays can produce narrow spot beams, the frequency can be re-used within the cell as long as the spatial separation between the users is sufficient. Figure 2.4 shows three users served by SDMA using the same channel within the cell. In a practical cellular environment it is improbable to have just one transmitter fall within the receiver beam width. Therefore it becomes imperative to use other multiple access techniques in conjunction with SDMA. When different areas are covered by the antenna beam, frequency can be re-used, in which case TDMA or CDMA is employed, for different frequencies FDMA can be used. SDMA architecture can be configured and deployed for most of the well-known mobile communication architectures such as CDMA (Code Division Multiple Access), TDMA (Time Division Multiple Access) and FDMA (Frequency Division Multiple Access).

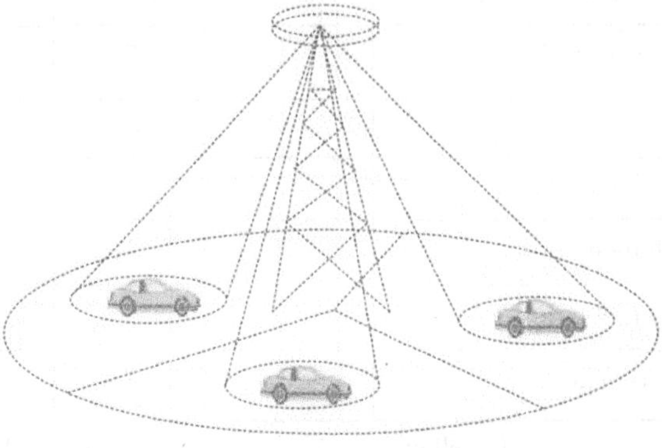

Fig. 2.4. Intra-cell SDMA

Workings of SDMA

SDMA architecture uses spatial multiplex technology that works through the application of simultaneous transmissions of radiation streams at the same frequency from the transmitter to the receiver both of which are equipped with multiple antennas. Using SDMA, both the transmitter and the receiver signal-processing ensures complete separation of signals in the spatial (or antenna) dimension.

The spatial multiplexing results in augmenting the bit rate without consuming extra time or frequency resources. The spatial multiplexing facilitates achieving higher throughput by sending several bit streams to one multi-antenna terminal without expending additional transmission power.

Using SDMA, the parallel spatial transmission of bit streams increases the wireless link capacity linearly with minimum transmission input, depending upon the number of antennas deployed at the receiver and the transmitter end.

Features

In a mobile cellular communication network, the SDMA leverages the spatial location of mobile terminals, equipments and devices within that cell, thereby enhancing the efficiency in network bandwidth utilization.

Unlike traditional mobile cellular network systems, where the base station is tied up, radiating radio signals in all directions within the cell, with no knowledge of the location of mobile devices, SDMA architecture enables the channeling of radio signals based on the mobile devices' locations.

In this way, SDMA architecture not only protects the quality of radio signals, safeguarding against interference causing noise and signal degradation coming from adjacent cells, but also saves on redundant signal transmission in areas where mobile devices are not currently active or unavailable.

Technology Comparison

Using five different performance metrics, the following table shows the differences between each of the technologies. Each technology is rated based on their performance with respect to the ideal performance level.

Performance parameters definitions given below:

Performance Parameters	*FDMA*	*TDMA*	*CDMA*
Capacity (Spectrum Efficiency)	1	2	3
Security	1	1	3
Ease of Network Planning	1	1	2
Ease of Implementation	4	4	2
Cost of Implementation	3	3	2
Backwards Compatibility	4	3	1
Excellent 4 3 2 1 0 Poor			

Performance parameters definitions given follows:

Capacity (Spectrum Efficiency): Measures the ability to handle heavy traffic.

Security: Measures the ability to keep information from being intercepted by others.

Ease of Network Planning: Measures the ease of creating and planning network structures.

Ease of Implementation: Measures the ease of carrying out a network system and equipment.

Cost of Implementation: Measures the financial requirements of carrying out a network system and equipment.

Backward Compatibility: Measures the ability to comply with existing systems.

2.2 PACKET RADIO MEDIA ACCESS PROTOCOLS

A medium access control protocol is needed for a multi-access network. Such protocols are designed to address the problem of how to share a common broadcast channel. The techniques that have been used for packet radio networks can be termed *random access* or *contention* techniques. They are random access in the sense that there is no predictable or scheduled time for any station to transmit: Stations generate packets for transmission at random times. They are contention in the sense that no central control is exercised to determine whose turn it is. All stations contend for time on a network.

There are three medium access protocols appropriate for packet radio networks: ALOHA, slotted ALOHA, and Carrier Sense Multiple Access (CSMA).

The maximum channel utilization is only 0.37 with slotted ALOHA .The reason for this poor performance is that both ALOHA and slotted ALOHA fail to take advantage of one of the key properties of packet radio networks, which is that *the propagation delay between stations is insignificant compared to packet transmission time.*

If the station-to-station propagation time is extremely small compared to packet transmission time, then, when a station launches a packet, all the other stations know it almost immediately. So, if they had any sense, they would not try transmitting until the first station was done. In that case, collision would be rare since they would occur only if two stations began to transmit almost simultaneously. In other word the short delay time provides the stations with better feedback about the state of the system.

This observation led to the development of a technique known as Carrier Sense Multiple Accesses (CSMA) or Listen Before Talk (LBT). A station wishing to transmit first listens to the medium to determine if another transmission is in progress. If the medium is in use, the station backs off some period of time and tries again. If the medium is idle, the station may transmit. To avoid collisions, a station waits a reasonable amount of time after transmitting for an acknowledgment, taking into account the maximum round-trip propagation delay. If there is no acknowledgment, the station assumes that a collision has occurred and retransmits.

With CSMA, an algorithm is needed to specify what a station should do if the medium is found to be busy. The following three algorithms are in use:

- Non-persistent CSMA
- Persistent CSMA
- P-persistent CSMA.

2.3 RESERVATION PROTOCOL

The RSVP protocol is part of a larger effort to enhance the current Internet architecture with support for *Quality of Service* flows. The RSVP protocol is used by a host to request specific qualities of service from the network for particular application data streams or flows. RSVP is also used by routers to deliver Quality of Service (QoS) requests to all nodes along the path(s) of the flows and to establish and maintain state to provide the requested service. RSVP requests will generally result in resources being reserved in each node along the data path.

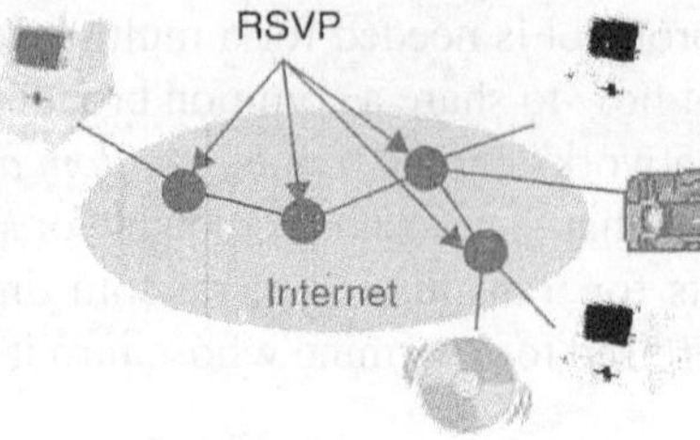

Fig. 2.5

2.3.1 RSVP Protocol Overview

A host uses RSVP to request a specific Quality of Service (QoS) from the network, on behalf of an application data stream. RSVP carries the request through the network, visiting each node the network uses to carry the stream. At each node, RSVP attempts to make a resource reservation for the stream.

To make a resource reservation at a node, the RSVP daemon communicates with two local decision modules, *admission control* and *policy control.* Admission control determines whether the node has sufficient available resources to supply the requested QoS. Policy control determines whether the user has administrative permission to make the reservation. If either check fails, the RSVP program returns an error notification to the application process that originated the request. If both checks succeed, the RSVP daemon sets parameters in a *packet classifier* and *packet scheduler* to obtain the desired QoS. The packet classifier determines the QoS class for each packet and the scheduler orders packet transmission to achieve the promised QoS for each stream.

A primary feature of RSVP is its scalability. RSVP scales to very large multicast groups because it uses receiver-oriented reservation requests that merge as they progress up the multicast tree. The reservation for a single receiver does not need to travel to the source of a multicast tree; rather it travels only until it reaches a reserved branch of the tree. While the RSVP protocol is designed specifically for multicast applications, it may also make unicast reservations.

RSVP is also designed to utilize the robustness of current Internet routing algorithms. RSVP does not perform its own routing; instead it uses underlying routing protocols to determine where it should carry reservation requests. As routing changes paths to adapt to topology changes, RSVP adapts its reservation to the new paths wherever reservations are in place. This modularity does not rule out RSVP from using other routing services. Current research within the RSVP project is focusing on designing RSVP to use routing services that provide alternate paths and fixed paths.

RSVP runs over IP, both IPv4 and IPv6. Among RSVP's other features, it provides opaque transport of traffic control and policy control messages, and provides transparent operation through non-supporting regions.

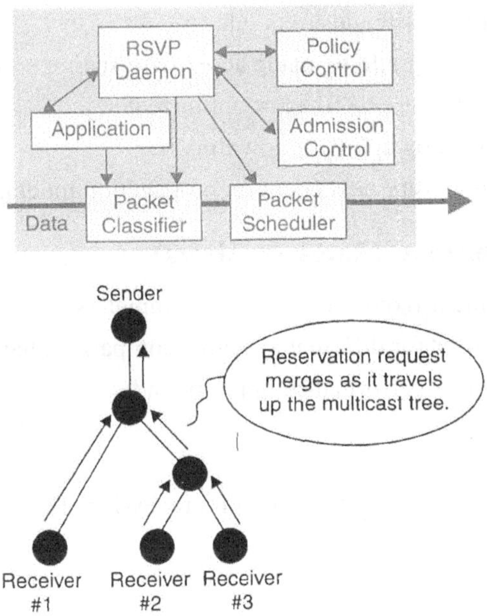

Fig. 2.6

2.4 CELLULAR SYSTEM CAPACITY

Shannon Capacity

- Maximum achievable rate or set of rates with arbitrarily small probability of error
- Coding scheme not achievable, and complexity/delay are infinite
- Tractable formulas exist for point-to-point links, MAC and degraded broadcast channels, and ad-hoc networks under various assumptions.

Other Theoretical Capacity Definitions

- Outage capacity
- Computation cutoff rate (not very useful).

Capacity Definitions Used in Practice

- Achievable rates under practical system assumptions
- User capacity under practical system assumptions.

Shannon-Theoretic Definition

- Multi-user channels typically assume user coordination and joint encoding/decoding strategies
- Can an optimal coding strategy be found, or should one be assumed (*i.e.,* TD, FD, or CD)?
- What base station(s) should users talk to?
- What assumptions should be made about base station coordination?
- Should frequency reuse be fixed or optimized
- Is capacity defined by uplink or downlink?
- Capacity becomes very dependent on propagation model.

Practical Capacity Definitions (Rates or Users)

- Typically assume a fixed set of system parameters
- Assumptions differ for different systems: comparison hard
- Does not provide a performance upper bound.

Shannon Capacity

- TDMA systems with joint base station processing (Hrish).

Multicell Capacity

- Rate region per unit area per cell
- Achievable rates determined via Shannon-theoretic analysis or for practical schemes/constraints
- Area spectral efficiency is sum of rates per cell.

User Capacity

- Calculates how many users can be supported for a given performance specification.
- Results highly dependent on traffic, voice activity, and propagation models.
- Can be improved through interference reduction techniques. (Gilhousen et. al.)
- Maximum number of users a cellular system can support in any cell.
- Can be defined for any system.
- Typically assumes symmetric data rates, cells, propagation, and mobility.
- Depends on the user specifications and radio design
- Data rate, BER, modulation, coding. etc.

Multicell Capacity

- Multi-user rate region per Hertz divided by coverage area given reuse distance

$$C_{\text{multicell}} = \frac{(R_1, ..., R_n)/B}{\pi(.5R_D)^2}$$

- Rate region $(R_1, \ldots, R_n)$ can be obtained via Shannon analysis
- How to treat interference from other cells
- Alternatively, can compute under practical system assumptions
- ASE sums rates in each cell:

$$A = \sum_i R_i/(B\pi(.5R_D)^2)$$

Capacity Calculation

- Calculate probability E_b/N_0 below target (BER exceeds target) based on Ns and these statistics.

$$P(\text{BER} > 10^{-3}) = P\left(\sum_{i=1}^{N} \chi_i + I/S > \delta\right)$$

$$\delta = \frac{W/R}{E_b/N_0} - \frac{\eta}{S}$$

- Compute outage probability as a function of N_s.
- Assumes target $E_b/N_0 = 5 \Rightarrow d = 30$
- Results indicate 60 users/sector with 1% outage.

UNIT **3** *WIRELESS NETWORKING*

3.1 INTRODUCTION

Whether it's because you have made a call using a mobile phone, received a message on your pager, checked your email from a PDA or even just seen an advert related to it, we have all come across a wireless data or voice network.

If a user, application or company wishes to make data portable, mobile and accessible then wireless networking is the answer. A wireless networking system would rid of the downtime you would normally have in a wired network due to cable problems. It would also save time and money due to the fact that you would spare the expense of installing a lot of cables. Also, if a client computer needs to relocate to another part of the office then all you need to do is move the machine with the wireless network card.

Wireless networking can prove to be very useful in public places—libraries, guest houses, hotels, cafeterias, and schools are all places where one might find wireless access to the Internet. From a financial point of view, this is beneficial to both the provider and the client. The provider would offer the service for a charge—probably on a pay per use system, and the client would be able to take advantage of this service in a convenient location; away from the office or home. A drawback of wireless Internet is that the QoS (Quality of Service) is not guaranteed and if there is any interference with the link then the connection may be dropped.

3.2 DIFFERENCES IN WIRELESS AND FIXED TELEPHONE NETWORKS

Fixed wireless access is the use of wireless technology to replace copper to connect subscribers to the telephone network. There are numerous acronyms for the use of wireless technology in local loop applications, as well as a variety of applications which require differences in the choice of wireless technology to optimize the economics, performance and spectrum utilization for each one.

The ability of different wireless technologies to co-exist within a single geographic area, owned and operated by different service providers (with relatively simple sharing arrangements for common infrastructure) is one of the key aspects which makes wireless technology a valuable addition to the Universal Service concept—in fact it can change the fundamental assumptions embodied within traditional approaches to Universal Service.

At the same time that new developments in wireless technology are providing more choices, regulators are in the process of redefining everything that is critical to local loop and access investment, for example :-

- Permitting the use of Cellular / PCS spectrum for Fixed Access
- Assignment of Spectrum for MMDS, LMDS and point-to-multipoint services
- Facilities Unbundling, Interconnection, Resale and Access Competition
- Access Charge reform
- Universal Service re-definition and subsidy reform.

There is a significant danger that these latter changes will be finalized around traditional (cabled) technology assumptions, without making full allowance for the new range of freedoms and choices enabled by wireless technology. This is therefore a timely opportunity to understand these issues within the context of Universal Service. I will concentrate on the technology and spectrum factors and in particular I will describe the new generation of Fixed Wireless Access (FWA) systems which are already being deployed in the rest of the world by three key North American vendors.

Table 3.1. Wireline Equivalence Technical Comparisons

Parameter	*Wireline Technologies*	*Fixed Wireless Technologies*	*Mobile Wireless Technologies*
Voice Coding	64 kb/s PCM or analog	32 kb/s ADPCM or 64kb/s PCM	4–16 kb/s compressed coding
Voice Quality	MOS = 4.3*	MOS = 4.0 – 4.3*	MOS = 3.5 – 4.1*
Group 3 Fax Rates	All to 14.4, or 28.8/ 33.6k when available	Same as wireline	2.4, 4.8, 9.6 or 14.4k (19.2k future)
Modem Data Rates	All to 33.6k	Same as wireline	2.4, 4.8, 9.6 or 14.4k (19.2k future)
Digital Data Rates	All to 56/64, 128, 384k	Same as wireline	9.6, 14.4k or 19.2k or Packet Mode
T1/E1 Delivery	Fractional or Full	Same as wireline	Not Practicable

Customer Interfaces	¡ RJ-11 or equivalent ¡ ISDN BRI ¡ Leased Line Standards ¡ Data Standards	Same as wireline	RJ-11 and data interface emulations via adapter boxes on AMPS, PCS or GSM air interfaces
Engineered Traffic Capacity per line	1–36 ccs	Same as wireline	1–10 ccs typical
Economic Access Grade of Service	Better than 1% or 0.1% blocking	Same as wireline	Typically 2–20% blocking
Long term Bit Error Rates	10^{-6}, 10^{-8}	Same as wireline	10^{-3}, 10^{-5}
One-way Loop Delay	<5 milliseconds	<20 milliseconds	50–150 milliseconds
• MOS scale = 1–4.5 basis (MOS = Mean Opinion Score)			

3.3 FIXED WIRELESS ACCESS TECHNOLOGIES AND STANDARDS

Table 3.2 below summarizes the three main classes of wireless access technologies. The first two are characterized by their need for rigidly standardized air interfaces in order to support mobility / portability and compatibility with retail terminals. The "Application Specific" class is characterized by the need to support standardized (fixed) interfaces at each end of the system, with non-standardized air interfaces optimized for the different services, capacity and applications supported.

Table 3.2. Wireless Access Technologies

	Fixed Cellular and PCS	*Cordless*	*Application Specific*
Original Application	Outdoor Mobile and Wide Area Portable Telephony	Indoor, campus, PBX and local portability	MMDS LMDS FWA Point-to-Point Point-to-Multipoint Satellite
Wireless Access Application	Low-tier POTS (esp Developing Countries)	Wireline Extension and portability	Full and Advanced POTS, business, TV and broadband. Universal Service

Services Voice	Compressed	Compressed or 32k	32k or 64k
Fax	Limited Speeds	Limited Speeds	Full Speed
Modem Data	Limited Speeds	Limited Speeds	Full Speed
Digital Data	Limited Speeds	Limited Speeds	56kb/s–45Mb/s
ISDN	No	No	BRI and PRI
Internet Access and LAN Bridge	Limited Speeds	Limited Speeds	28.8kb/s–10Mb/s
Standards	AMPS (FDMA) IS-54 and IS-136(TDMA) IS-95 (CDMA) NMT (FDMA) GSM (CDMA)	CT0/1 (Analog) CT2 (TDMA) PACS (TDMA) PHS (TDMA) DECT (TDMA)	Vendor Specific
Frequency	400 / 800 / 900 MHz 1800 / 1900 MHz Licensed / auctioned	45 MHz 860 MHz 1800 MHz Often unlicensed	Various frequencies Mostly licensed Some auction or fee-based charging
Typical Range City	1–2 miles	500–1000 feet	1–2 miles
Urban	2–5 miles	1000–2000 feet	2–5 miles
Suburban	5–10 miles	1–2 miles	5–10 miles
Rural	10–30 miles	2–3 miles	10–30 miles
Network Elements Type	Centralized Overlay with own Controller	Cluster/Zone Controllers	Distributed
Switch and Interface	Mobile/Trunks	POTS or BRI	Various
Terminals	Fixed/Mobile/Portable	Fixed/Portable	Fixed/Portable
Base Station	Large Cell Site	Pole/Building Mounted	Varies
Base Station Controller	Yes	Yes	Generally None

How will fixed wireless service and mobile wireless services be differentiated?

There are actually four services implied here:

- The mobile service offered by cellular and PCS operators, directly or via resale
- Any fixed (or hybrid fixed / mobile / portable) options based on these mobile technologies and networks. These services do not support the same feature, functionality or quality criteria as fixed wireline networks, and are not subject to the same regulatory and service rules.

- Fixed wireless installations based on cellular / PCS technologies but connected to fixed network switches and services. These are almost Wireline Equivalent but often cannot transparently support all the features, functions or quality aspects of wireline access.
- Wireline Equivalent FWA. This appears to the user and the service provider as a transparent equivalent to wireline service, and does not need to be differentiated.

3.4 LIST OF ACRONYMS

ADPCM Adaptive Pulse Code Modulation
ADSL Asynchronous Digital Subscriber Line
AMPS Advanced Mobile Phone System/Service
ANI Automatic Number Identification
CATV Community Antenna Television
CITEL Inter-American Telecommunications Commission
CITI Columbia Institute for Tele-Information
CPE Customer Premise Equipment
DTMF Dual Tone Multi-Frequency
FDMA Frequency Division Multiple Access
FWA Fixed Wireless Access
GSM General Service Mobile
ISDN Integrated Services Digital Network
ISDN BRI ISDN Basic Rate Interface
ISDN PRI ISDN Primary Rate Interface
LAN Local Area Network ,LEC Local Exchange Carrier
LMDS Local Multipoint Distribution Service
LNP Local Number Portability
MMDS Multichannel Multipoint Distribution Service
MOS Mean Opinion Score
OA and M Operations, Administration and Maintenance
OAM and P Operations, Administration, Maintenance and Planning
PACS Personal Access Communications System
PBX Private Branch Exchange
PCM Pulse Code Modulation
PCS Personal Communications Service
PHS Personal Handi-phone System

POTS Plain Old Telephone Service

PSTN Public Switched Telephone Network

SOHO Small Office Home Office

SONET Synchronous Optical Network

SS#7 Signaling System Number 7.

3.5 PACKET SWITCHING

Packet switching refers to protocols in which messages are broken up into small packets before they are sent. Each packet is transmitted individually across the net. The packets may even follow different routes to the destination, depends on the type of packet switching. Thus, each packet has header information in which enable to route the packet to its destination. At the destination the packets are reassembled into the original message.

To prevent unpredictably long delays and ensure that the network has a reliably fast transit time, a maximum length is allowed for each packet. It is for this reason that a message submitted to the transport layer may first have to be divided by the transport protocol entity into a number of smaller packet units before transmission. In turn, they will be reassembled into a single message at the destination.

3.5.1 Circuit Switching vs. Packet Switching

In Circuit switching networks, when establishing a call a set of resources is allocated for this call. These resources are dedicated for this call, and can' be used by any of the other calls. Circuit Switching is ideal when data must be transmitted quickly, must arrive in sequencing order and at a constant arrival rate. There for when transmitting real time data, such as audio and video, Circuit Switching networks will be used.

Packet switching main difference from Circuit Switching is that the communication lines are not dedicated to passing messages from the source to the destination. In Packet Switching, different messages can use the same network resources within the same time period. Since network resources are not dedicated to a certain session the protocol avoid from waste of resources when no data is transmitted in the session. Packet Switching is more efficient and robust for data that is burst in its nature, and can withstand delays in transmission, such as e-mail messages, and Web pages.

3.5.2 Two Kinds of Packet Switching

There are two basic types of packet switching:

1. *Virtual Circuit Packet Switching Networks*

   An initial phase is used to set up a route between the intermediate nodes for all the packets passed during the session between the two end nodes. In each intermediate node, an entry is registered in a table to indicate the route for the connection that has been set up. The packets passed through this route, have short headers, containing only a Virtual Circuit Identifier (VCI). Each intermediate node passes the packets according to the information that was stored in its table, in the set up phase and according to the packets header content.

In this way, packets arrive at the destination in the correct sequence. This approach is slower than Circuit Switching, since different virtual circuits may compete over the same resources. As in Circuit Switching, if an intermediate node fails, all virtual circuits that pass through it are lost.

The most common forms of Virtual Circuit networks are ATM and Frame Relay, which are commonly used for Public Data Networks (PDN).

2. ***Datagram Packet Switching Networks***

   This approach uses a different, more dynamic scheme, to determine the route through the network links. Each packet is treated as an independent entity, and its header contains full information about the destination of the packet. The intermediate nodes examine the header of the packet, and decide the next hop of this packet. In the decision two factors are taken into account:

   - The shortest way to pass the packet to its destination—protocols such as RIP/OSPF is used to determine the shortest path to the destination.
   - Finding a free node to pass the packet to—in this way, bottle necks are eliminated, since packets can reach the destination in alternate routes. Thus, in this method, the packets don't follow a pre-established route, and the intermediate nodes (the routers) don't have pre-defined knowledge of the routes that the packets should be passed through.

Packets can follow different routes to the destination. Due to the nature of this method, the packets can reach the destination in a different order than they were sent, thus they must be sorted at the destination to form the original message. This approach is time consuming since every router has to decide where to send each packet.

The main implementation of Datagram Switching network is the Internet which uses the IP network protocol.

3.6 SWITCH FUNCTIONALITY

The world of broadband data communications is based on interconnecting small networks to larger ones. At the lowest level we always find fast hardware-based switches, capable of delivering information from the various interconnected networks to their correct destinations. Because of the high data rates involved, routing decision-making logic must be implemented entirely in hardware; software processing would be grossly incapable of keeping pace with the high rate of packet arrivals. The implementation normally requires custom-design VLSI circuitry to maintain physical compactness. In this discussion, we will assume that the arriving packets are of fixed length. Such fixed-length packets are also referred to as *cells.* Two primary functions are supported by the N-input, N-output switch (subsequently referred to as an NN switch):

1. Routing of each cell to its correct destination port.

2. Resolving the contention which arises when several simultaneously arriving cells are headed to a common output port. This is solved by means of store-and-forward buffering: arriving packets which cannot be immediately delivered are stored in a buffer (or buffers) within the switch, their delivery being deferred according to a buffer servicing discipline.

3.7 INTRODUCTION TO X.25

The use of this protocol started in the 1970's. X.25 protocol started its way due to the need to provide users with wide area network (WAN) connectivity across public data networks (PDNs). It was felt that improved equipment compatibility and lower cost would increase subscription to PDNs. When started its way, the X.25 protocol was the most popular one among others that were developed at that time.

X.25 was developed by the common carriers, rather than any single commercial enterprise. Because of that, the specifications of the protocol are well suited to various system types.

One of the big advantages of the X.25 is the fact that its carry an international nature. X.25 is managed by the an agency of the United Nations called the International Telecommunications Union (ITU). That's why X.25 is known as a global standard.

What is an X.25 Network?

An X.25 network provides a means by which one X.25 DTE (a Terminal or Host of some kind) can exchange data with one or more other X.25 Host, on the other side of the network.

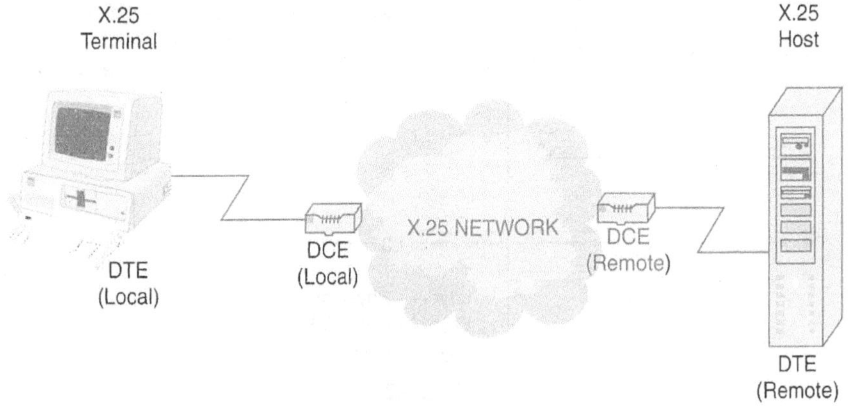

Fig. 3.1

Data is carried within individual packets—X.25 is often referred to as a Packet Switching Protocol. This makes it similar to a TCP/IP network—the difference is that IP networks employ a Connectionless protocol: each packet is routed according to the information within that packet (typically by using the Destination Address). By contrast, X.25 is a Connection-Oriented protocol: the routing information used by the network is carried only in the packets used to establish the connection;

thereafter addressing information is not required. This does, however, mean that the X.25 network switching nodes need be aware of each connection, unlike IP routers.

X.25 Layers

The X.25 protocol is divided into 3 layers (or Levels); TCP/IP, on the other hand is divided into 4 layers.

Layer	*Description*	*X.25*	*TCP/IP Equivalence*
1	Physical Layer	V.24, X.21, V.35, etc	Same as X.25, and others not normally used by X.25 (too many to mention)
2	Data Link Layer	LAPB	Many, such as PPP, SLIP, Cisco HDLC, etc and even X.25 itself!
3	Network Layer	X.25 PLP	IP
4	Transport Layer	(none)	TCP or UDP

Each of these layers is independent. The Physical layer includes the mechanical and electrical aspect of communications—in other words, cabling. The X.25 Data Link Layer provides the reliable link between the DTE and the DCE (or Network), and the X.25 Packet Layer Protocol (PLP) provides the information necessary to make and maintain a connection across the network.

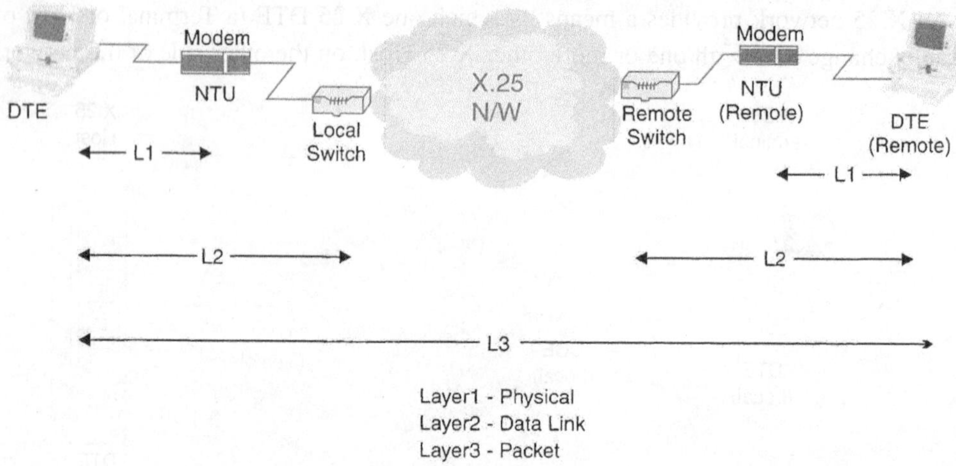

Fig. 3.2

It's probably most useful to think of Layer 1 as being the physical connection to the network NTU or modem, Layer 2 as being the Logical Link between the DTE and the local network switching node, and a Layer 3 Virtual Circuit as being the connection to the remote DTE.

X.25 Addresses

An X.25 Network User Address (NUA) is much like a telephone number, being a string of digits, and can be up to 15 digits in length. The NUA on a typical network will be 12 digits in length, with another 2 digits for the subaddress.

Advantages/Disadvantages

Advantages

The main benefit that X.25 provides is capability of reliable data communications over a wide range of quality for transmission facilities. X.25 can operate over either digital or analog facilities while frame relay and ATM technologies depend on the use of high quality digital transmission facilities.

Another benefit of X.25 is due to the seniority of this protocol. X.25 has been created in the mid 1970's so it has seniority of over 30 years. Consequently, it is stable and debugged very well. One can't find any data errors when using modern X.25 networks what makes it very reliable protocol.

X.25 gives a virtual high quality digital network at low cost. It is economical for the reason that there are tremendous savings to be made if multiple parties share the same infrastructure. X.25 is usually paid for by a monthly connect fee plus packet charges. There is usually no holding charge; therefore X.25 is suitable for organizations that must be on line all the time.

Another feature which is very useful is X.25 capability of speed matching: because of the store-and-forward nature of packet switching and the very good flow control, data transfer equipments do not have to use the same line speed. For example, you can have a host connected at 64kbps communicating with several remote sites connected with 19kbps lines.

Disadvantages

The major disadvantages of X.25 that causes it to be less useful in today's protocols world are the following:

First of all, it can be very slow, a feature that makes it unusable for most of applications. For example, it's terrible for telephony, because the human ear can't stand such delays. The store-and-forward mechanism of X.25 caused it to have an inherent delay. Frame Relay, on the contrary, reduces the transmission delay significantly since it does not store and forward, but simply switches to the destination.

Another problem of using X.25 protocol is the fact that it requires a massive buffering capability in order to support the store-and-forward mechanism of data transfer which makes it very expensive and not efficient to use. One of the main reasons that Frame Relay is relatively so cheap is that its requirements of storage are minimal.

UNIT 4 WIRELESS DATA SERVICES

4.1 INTRODUCTION

The wireless Internet-enabled subscriber base continues to grow. Telecommunications companies and wireless service providers are looking for ways to rapidly create and deploy new revenue-generating Internet services while providing their wireless subscribers with a more personalized experience. Because wireless subscribers have a different set of essential desires and needs than desktop or even laptop Internet users, Internet services must be designed to present the optimum experience for subscribers using different types of wireless devices. Writing Internet services in this manner requires a deep understanding of the technical issues unique to the wireless environment.

WAP sessions cope with intermittent coverage and can operate over a wide variety of wireless bearer networks including but not limited to Cellular Digital Packet Data (CDPD), Code Division Multiple Access (CDMA) and Global System for Mobiles (GSM), Time Division Multiple Access (TDMA) and General Packet Radio Service (GPRS). Although most WAP services in Europe and the United States are circuit-switched (dial-up), WAP will also work on packet-switched networks.

4.2 CELLULAR DIGITAL PACKET DATA (CDPD)

Cellular Digital Packet Data (CDPD) uses unused bandwidth normally used by AMPS mobile phones between 800 and 900 MHz to transfer data. Speeds up to 19.2 kbit/s are possible. Developed in the early 1990s, CDPD was large on the horizon as a future technology. However, it had difficulty competing against existing slower but less expensive Mobitex and DataTac systems, and never quite gained widespread acceptance before newer, faster standards such as GPRS became dominant.

CDPD had very limited consumer offerings. AT and T Wireless first offered the technology in the United States under the PocketNet brand. It was one of the first consumer offerings of wireless web service. A company named Omnisky provided service for Palm V devices. Cingular Wireless later offered CDPD under the Wireless Internet brand (not to be confused with Wireless Internet Express, their brand for GPRS/EDGE data). PocketNet was generally considered a failure with competition from 2G services such as Sprint's Wireless Web. After the four phones AT and T Wireless had offered to the public (two from Panasonic, one from Mitsubishi and the Ericsson R289LX), AT and T Wireless eventually refused to activate the devices.

Despite its limited success as a consumer offering, CDPD was adopted in a number of enterprise and government networks. It was particularly popular as a first-generation wireless data solution for telemetry devices (machine to machine communications) and for public safety mobile data terminals.

In 2004, major carriers in the United States announced plans to shut down CDPD service. In July 2005, the AT and T Wireless and Cingular Wireless CDPD networks were shut down.

Radio Data System

Radio Data System, or **RDS,** is a communications protocol standard for embedding small amounts of digital information in conventional FM radio broadcasts. The RDS system standardizes several types of information transmitted, including time, station identification and programme information.

Radio Broadcast Data System (RBDS) is the official name used for the U.S. version of RDS. The two standards are nearly identical, with only slight differences.

The standard began as a project of the European Broadcasting Union (EBU), but has since become an international standard of the International Electrotechnical Commission (IEC).Both use a 57 kHz subcarrier to carry data at 1187.5 bits per second. The 57 kHz frequency was chosen for being the third harmonic of the pilot tone for FM stereo, so it would not cause interference or intermodulation with it, or with the stereo difference signal at 38 kHz (the second harmonic). The data format utilizes error correction. RDS defines many features including how private (in-house) or other undefined features can be "packaged" in unused program groups.

4.3 ADVANCED RADIO DATA INFORMATION SERVICES (ARDIS)

Advanced Radio Data Information Services (ARDIS) is a two-way radio service that is based on Motorola's RD-LAP technology. It was originally created and jointly owned by Motorola and IBM to serve IBM field technicians. In 1995, Motorola acquired 100 percent ownership of ARDIS and in 1998; ARDIS was acquired by the American Mobile Satellite Corporation. ARDIS is quoted as covering 90 percent of the urban business population and more than 400 metropolitan areas in the United States, Puerto Rico, and the Virgin Islands. It is estimated that it has more than 40,000 users, and many of them are still IBM customer engineers.

Motorola has developed two proprietary air-interface protocols for the ARDIS packet network: the MDC-4800, which provides 4,800 b/s services, and the RD-LAP, which provides 19,200 b/s services. MDC-4800 service has been deployed throughout the ARDIS network and only a few major service areas have been enhanced with RD-LAP capability.

ARDIS co-founded the Worldwide Wireless Data Network Operators Group (WWDNOG), an international forum for wireless data network service providers, which includes network operators in Canada, the United Kingdom, Germany, Hong Kong, Australia, Malaysia, and Thailand. The group facilitates the development of common software applications for RD-LAP technology and enables interconnectivity among its member network.

4.3.1 System Description

ARDIS' network is deployed through base stations that are connected back to radio network controllers via dedicated leased lines. ARDIS has approximately 1400 base stations occupying 1100 cell sites. Remote users access the system from laptop radio terminals, which communicate with the base stations. Each of the ARDIS base stations is tied to one of the several radio network controllers, as shown in Fig. 4.1. A radio network controller is a specialized computer also called a Radio Frequency/Network Control Processor (RF/NCP). The RF/NCP manages the RF resources, including the radio base stations and data sent over the radio channels. Both inbound and outbound channels are managed using totally different delivery strategies. The RF/NCP evaluates the strength of the signal received from every wireless device transmission at each base station for each inbound data packet that it "hears." It then selects the best base station to communicate with that particular wireless device and will send the next outbound message through that base station. This process ensures the optimum selection of a site, although subscriber devices may be moving or be subject to RF noise.

The RF/NCP is one network component that helps manage the automated roaming capability of the ARDIS network. Wireless devices can automatically move (roam) between any of the ARDIS frequencies on either of the two air protocols (MDC-4800 or RD-LAP 19.2), or between any of the ARDIS layers that have been configured for in-building or on-street usage.

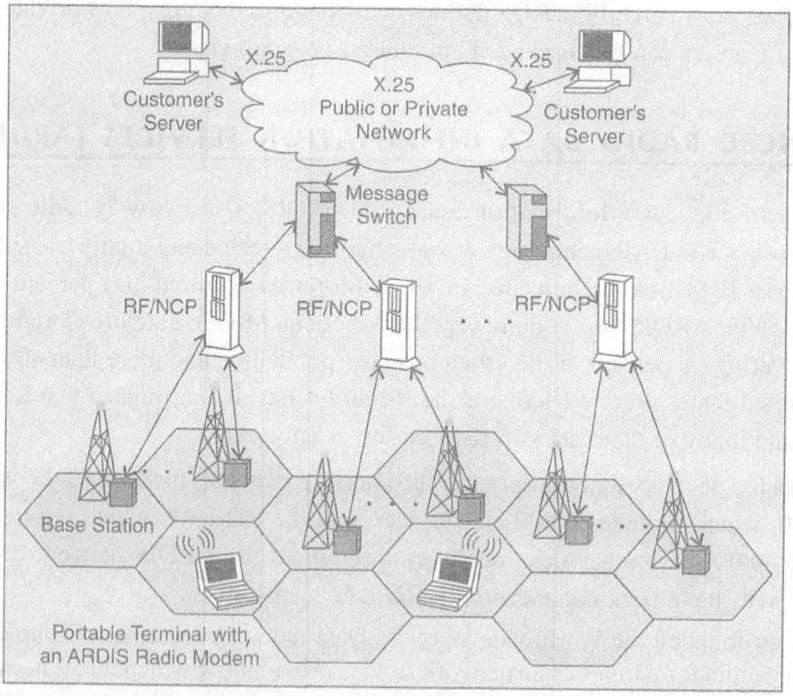

Fig. 4.1. ARDIS network architecture

The ARDIS radio terminals access the network using a random access method called Data Sense Multiple Accesses (DSMA). Before every transmission, a radio terminal listens to the base station transmitter to determine if a "busy bit" is on or off. This bit indicates whether the base station is currently receiving data (busy bit is on) or not (busy bit is off) and, effectively, it encodes the current state of the shared uplink channel. When this bit is on, radio terminals are prohibited from transmitting and, in that phase, the packet collisions are avoided. On the other hand, when the busy bit is off, a radio terminal is allowed to transmit. However, if two terminals happen to begin transmission at the same time, the packets collide and retransmission will be attempted, as in other contention-based multiple-access protocols.

4.3.2 Protocol Architecture

Figure 4.2 shows some of the key protocols used by ARDIS. Proprietary data exchange protocols are used to transport data at the various interface points .At the customer's server (or host), the Standard Context Routing (SCR) protocol is used to transport data between this server and the MS. The messages for transmission are broken into packets, each one with a header, containing routing and size information, and a data message. Host messages can be one of the following three basic types:

- Host Message Delivery Request (HR). used by the host to send messages to mobile terminals.
- Host Message Delivery Confirmation (HC), used by the MS to confirm message receipt.
- Mobile Device Message Delivery Indication (MI), used by the mobile to acknowledge message receipt.

In addition, extended SCR message types are used to monitor the mobile's activity, perform loopback diagnostics, and check status.

At the mobile terminal end, the Native Control Language (NCL) protocol is used as a link layer protocol between the client application on the mobile terminal and the radio modem. The NCL protocol provides a transaction-oriented service particularly suitable for connectionless applications. A command is issued from the application and a response is elicited from the radio modem for each command issued. Information is exchanged between the mobile terminal and the radio modem in packets called Service Data Units (SDUs). NCL provides three types of SDU:

- Command SDUs, which issue instructions such as "send" and "receive" messages.
- Response status SDUs, which are responses to commands.
- Event report SDUs, which can be enabled or disabled.

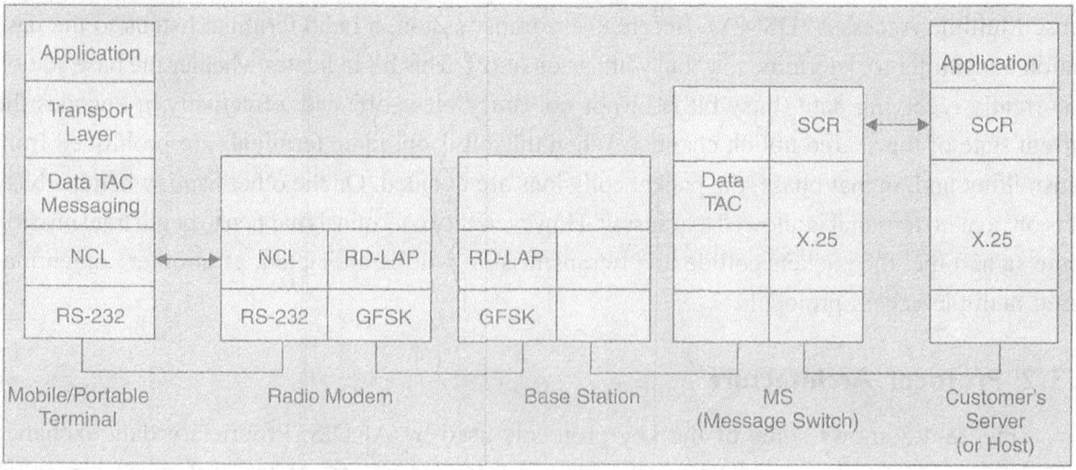

Fig. 4.2. Key protocols in ARDIS

4.4 RAM MOBILE DATA

RAM Mobile Data was originally founded by RAM Broadcasting Corporation as American Mobile Data Communications, Inc. in 1988 The name of the company was changed to RAM Mobile Data in 1989. RAM Mobile Data was the U.S. Operator of the Mobitex network.

RAM Mobile Data was sold and renamed BellSouth Wireless Data in 1995 and later became Cingular Interactive when BellSouth and SBC formed Cingular Wireless. The Mobitex division within Cingular Wireless was sold to an investment company in 2005 and became Velocita Wireless. Velocita Wireless was purchased by Sprint Nextel and became a Sprint Nextel Company in early 2006. Today RAM Mobile Data is the exclusive operator of Mobitex in the Netherlands. With HQ in Maarssen (NL) they are running the entire Mobitex operations throughout the BeNeLux.

RAM
MOBILE
DATA

Fig. 4.3

Navara, a division of RAM Mobile Data, provides mobile device interface solutions for existing applications and databases. Navara is an Independent Software Vendor (ISV) that produces the Navara Mobility Suite, a software system that allows mobile access to business systems. Navara is utilized to provide mobile access to many different backend systems across many industries. Navara support most major mobile devices and industry-standard integration and communication technologies.

4.5 COMMON CHANNEL SIGNALING

In telephony, **Common Channel Signaling (CCS)** or in the US **Common Channel Interoffice Signaling (CCIS)** is the transmission of signaling information (control information) on a separate channel to the data, and, more specifically, where that signaling channel controls multiple data channels.

For example, in the PSTN one channel of a communications link is typically used for the sole purpose of carrying signaling for establishment and clear down of telephone calls. The remaining channels are used entirely for the transmission of voice data. In most cases, a single 64kbit/s channel is sufficient to handle the call setup and call clear-down traffic for numerous voices and data channels. The logical alternative to CCS is Channel Associated Signaling (CAS), where each bearer channel has a signaling channel dedicated to it.

CCS offers the following advantages over CAS, in the context of the PSTN

- Faster call set up.
- No falsing interference between signaling tones by network and speech frequencies.
- Greater trunking efficiency due to the quicker set up and clear down, thereby reducing traffic on the network.
- No security issues related to the use of in-band signaling with CAS.
- CCS allows the transfer of additional information along with the signaling traffic providing features such as caller ID.

The most common CCS signaling methods in use today are ISDN and SS7.

ISDN signaling is used primarily on trunks connecting end-user PBX systems to a central office. SS7 is primarily used within the PSTN. The two signaling methods are very similar since they share a common heritage and in some cases, the same signaling messages are transmitted in both ISDN and SS7.CCS is distinct from in-band or out-of-band signaling, which are to the data band what CCS and CAS are to the channel.

4.6 ISDN (INTEGRATED SERVICE DIGITAL NETWORK)

ISDN stands for Integrated Service Digital Network, and as the name suggests it allows digital communication. This is favorable as digital technology is a lot faster, and more accurate than the old analogue lines as they no longer require the process of modulation and demodulation. ISDN relies on already existing copper cable systems, causing its integration into our existing communications system to be smoother and less disruptive.

According to ITU-T (formerly CCITT) "an ISDN is a network, in general evolving from telephony IDN, that provides end-to-end digital connectivity to support a wide range of services, including voice and non-voice services, to which users have addressed by a limited set of standard multi-purpose user-network interfaces." The key point of this definition is the ability to support voice services adequately; this has not been achieved using any other concept.

ISDN has four major aspects; telephone network; integrated services; digital; network. These will be explored in a little more detail later.

What is Broadband ISDN?

Narrowband ISDN has been designed to operate over the current communications infrastructure, which is heavily dependent on the copper cable. B-ISDN however, relies mainly on the evolution of fibre optics. According to CCITT B-ISDN is best described as 'a service requiring transmission channels capable of supporting rates greater than the primary rate.' Behind this statement lies the plan for a network and services that will have far more impact on the world we know today, than ISDN ever would.

ISDN is Telephone Network

When ISDN is referred to as a network it is to be considered a telephone network, not a computer network. Broadband ISDN allows its users to communicate over high speed, high quality digital channels. The media is supports include Telex, fax, voice telephone, video telephone, audio, high definition TV and computer networking.

ISDN is Integrated Services

In the past video, audio, voice and data services needed different types of communication channels. One of the main advantages of ISDN is the ability to integrate these features over the same network and cable plant. Not only is this possible using ISDN technology but the quality of the transmission is better also. In the past four networks were needed and video was distributed on coaxial lines, audio over balanced lines, voice used copper cable pairs and data services required coaxial or twisted pair cables. Using one network allows reductions in installation costs, as well as easier installation. Other features available include demand networking, automatic bandwidth and on the fly connectivity. Advances in the services available are due to ISDN being digital.

ISDN is Digital

Data applications, in particular, seemed to have problems with the old analogue services. This is due to the fact that computers are digital devices and the transmission of data needs to be modified form binary to analogy tones, and then changed back to binary when it is received. This process requires a modem, which handles the MODulation and DEModulation of the data. Whilst the data is in transit it is susceptible to outside influences like noise, line spikes and echoes. Bandwidth is also limited, with the speed of modems being close to the maximum possible.

ISDN is a Network

Networks require high speed connectivity if they are going to be useful. ISDN is an excellent vehicle for connecting LANs, because it scales in increments of 64 kilobits per second (kbps). Computers are digital devices, as is ISDN, meaning that no translation of information is required, improving quality and speed. Due to the characteristics of ISDN it can be used, with the same level of performance, across a room or halfway across the continent. This has unlimited benefits.

Speed of BISDN in Relation to Other Communications Mediums

The following diagram compares transfer rates of current digital communications technologies

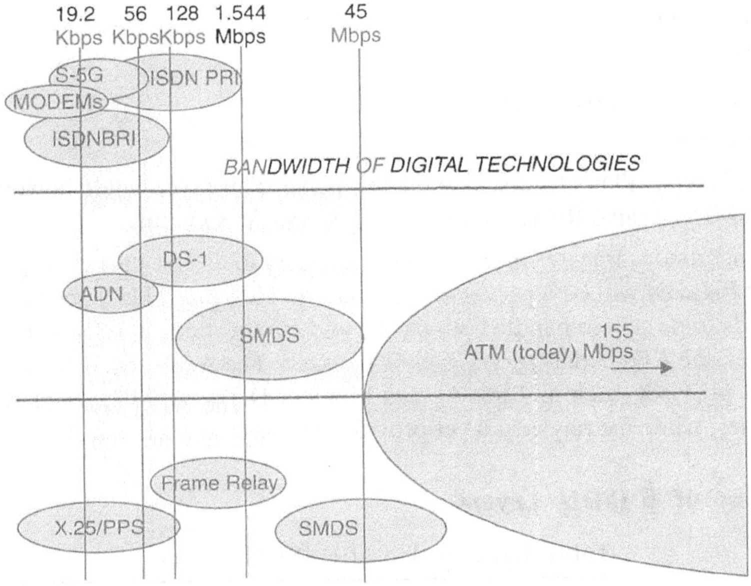

Fig. 4.4

4.6.1 Broadband ISDN Protocol

The suggested architecture for the B-ISDN protocol is depicted in figure 5 below. As can be seen the B-ISDN protocol uses a three plane approach. These three separate planes are referenced as:

- User Plane
- Control Plane
- Management Plane

The user plane is responsible for user information transfer including flow control and error control. The control plane manages the call-control and connection-control functions. While the management plane includes plane management, and layer management.

Figure: B-ISDN Protocol Reference Model

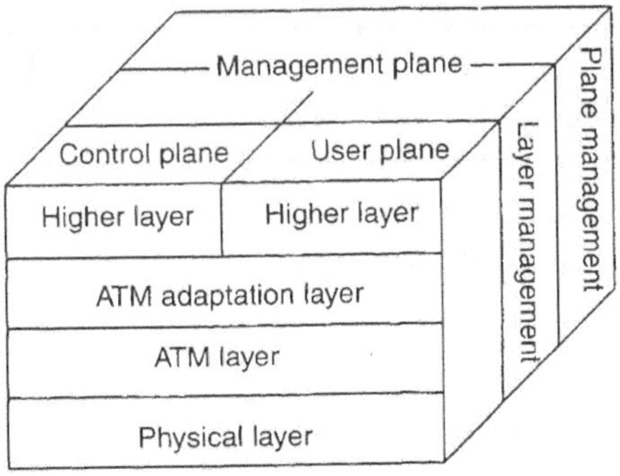

Fig. 4.5

The protocol for B-ISDN also adopts a layered approach, made up of four layers:

- Physical Layer
- ATM Layer
- ATM Adaptation Layer (AAL)
- Higher Layers

The higher layers will not be discussed in this paper. This layer would however be a services layer for video, SMDS, Frame Relay, and Access & Network Signaling.

ATM, asynchronous transfer mode, is often refereed to as fast packet switching. For this it can be taken that B-ISDN will be a packet based network. However ITU-T recommendations state that B-ISDN needs to be able to handle both packet and circuit-mode applications. Thus the use of an adaptation layer, the ATM Adaptation Layer, is required. The AAL will, therefore, be required to handle non-ATM protocol, such as Link Access Protocol-D.The ATM layer provides the packet transfer capabilities, while the physical layer provides the base network functions.

4.6.2 Functions of B-ISDN Layers

Table 4.1. Function of B-ISDN Layers

	Higher Layer Functions		*Higher Layers*
	Convergence	CS	
	Segmentation and reassemble	SAR	AAL
Layer Management	Generic flow control cell header generation/extraction Cell VPI/VCI translation Cell multiplex and demultiplex	ATM	
	Cell rate decoupling HEC header sequence generation/verification Cell delineation Transmission frame adaptation Transmission frame generation/recovery	TC	Physical Layer
	Bit timing Physical medium	PM	

CS = Convergence Sub Layer

SAR = Segmentation and Reassemble Sub Layer

AAL = ATM Adaptation Layer

ATM = Asynchronous Control Sub Layer

TC = Transmission Control Sub Layer

PM = Physical Medium Sub Layer.

4.7 ATM: ASYNCHRONOUS TRANSFER MODE PROTOCOL

The Asynchronous Transfer Mode (ATM) composes a protocol suite which establishes a mechanism to carry all traffic on a stream of fixed 53-byte packets (cells). A fixed-size packet can ensure that the switching and multiplexing function could be carried out quickly and easily. ATM is a connection-oriented technology, *i.e.;* two systems on the network should inform all intermediate switches about their service requirements and traffic parameters in order to establish communication.

The ATM reference model, which has two forms - one for the User-to-Network Interface (UNI) and the other for the Network-to-Node Interface (NNI), is divided into three layers: the ATM adaptation layer (AAL), the ATM layer, and the physical layer. The AAL interfaces the higher layer protocols to the ATM Layer, which relays ATM cells both from the upper layers to the ATM Layer and vice versa. When relaying information received from the higher layers, the AAL segments the data into ATM cells. When relaying information received from the ATM Layer, the AAL must reassemble the payloads into a format the higher layers can understand. This is called Segmentation and Reassembly (SAR). Different AALs are defined in supporting different types of traffic or service expected to be used on ATM networks.

The ATM layer is responsible for relaying cells from the AAL to the physical layer for transmission and from the physical layer to the AAL for use at the end systems, it determines where the incoming cells should be forwarded to, resets the corresponding connection identifiers and forwards the cells to the next link, as well as buffers cells, and handles various traffic management functions such as cell loss priority marking, congestion indication, and generic flow control access. It also monitors the transmission rate and conformance to the service contract (traffic policing).

The physical layer of ATM defines the bit timing and other characteristics for encoding and decoding the data into suitable electrical/optical waveforms for transmission and reception on the specific physical media used. In addition, it also provides frame adaptation function, which includes cell delineation, Header Error Check (HEC) generation and processing, performance monitoring, and payload rate matching of the different transport formats used at this layer. SONET, DS3, Fiber, twisted-pair are few media often used at the physical layer.

4.7.1 Protocol Structure of ATM

Each individual ATM cell consists of a 5-byte cell header and 48 bytes of information encapsulated within its payload. The ATM network uses the header to support the virtual path and the virtual channel routing and to perform a quick error check for corrupted cells.

The 5-byte header is structured as shown below:

```
|  8  |  7  |  6  |  5  |  4  |  3  |  2  |  1  |  bits
+-----+-----+-----+-----+-----+-----+-----+-----+
|   Generic Flow Control*   |  Virtual Path Identifier  |
+---------------------------+---------------------------+
|   Virtual Path Identifier |  Virtual Channel Identifier |
+---------------------------+---------------------------+
|              Virtual Channel Identifier               |
+---------------------------+-----------------+---------+
| Virtual Channel Identifier | Payload Type Id |   CLP   |
+---------------------------+-----------------+---------+
|                 Header Error Control                  |
+-------------------------------------------------------+
|                 Information Payload                   |
```

* For the UNI header only. The NNI header contains an additional 4 bytes of VPI information here instead.

H	GFC or VPI	VPI	
E	VPI	VCI	
A	VCI		
D	VCI	PT (3 Bit)	CLP
E	HEC		
R			
E	Cell Payload (48 Bytes)		

Fig. 4.6. ATM cell format

Generic Flow Control (GFC)

The GFC field of the header is only defined across the UNI. It is intended to control the traffic flow across the UNI and to alleviate short-term overload conditions. It is currently undefined and these 4 bits must be set to 0's.

Virtual Path Identifier (VPI)

The VPI, an 8-bit field for the UNI and a 12-bit field for the NNI, is used to identify virtual paths. In an idle cell, the VPI is set to all 0's. (Together with the Virtual Channel Identifier, the VPI provides a unique local identification for the transmission).

Virtual Channel Identifier (VCI)

This 16-bit field is used to identify a virtual channel. For idle cells, the VCI is set to all 0's. (Together with the Virtual Path Identifier, the VCI provides a unique local identification for the transmission).

Payload Type Identifier (PTI)

The three bits of the PTI are used for different purposes. Bit 4 is set to 1 to identify operation, administration, or maintenance cells (*i.e.,* anything other than data cells).

Bit 3 is set to 1 to indicate that congestion was experienced by a data cell in transmission and is only valid when bit 4 is set to 0.

Bit 2 is used by AAL 5 to identify the data as Type 0 (beginning of message, continuation of message; bit = 0) or Type 1 (end of message, single-cell message; bit = 1) when bit 4 is set to 0. It may also be used for management functions when bit 4 is set to 1. This bit is currently carried transparently through the network and has no meaning to the end user when AAL5 is NOT in use.

Cell Loss Priority (CLP)

The 1-bit CLP field is used for explicit indication of the priority of the cell. It may be set by the AAL Layer to indicate cells to discard in cases of congension, or by the network as part of the traffic management on commercial subscriber networks.

Header Error Control (HEC)

This is an 8-bit cyclical redundancy check computed for all fields of the first 4 bytes of the ATM cell header ONLY. It is capable of detecting all single-bit errors and some multiple-bit errors.

The HEC is compared by each switch as the ATM cell is received and all cells with HEC discrepancies (errors) are discarded. Cells with single-bit errors may be subject to error correction (if supported or discarded).

When a cell is passed through the switch and the VPI/VCI values are altered, the HEC is recalculated for the cell prior to being passed out the port.

4.8 SIGNALING SYSTEM #7 (SS7)

There are two essential components to all telephone calls. The first, and most obvious, is the actual content-our voices, faxes, modem data, etc. The second is the information that instructs telephone exchanges to establish connections and route the "content" to an appropriate destination. Telephony signaling is concerned with the creation of standards for the latter to achieve the former. These standards are known as protocols. SS7 or Signaling System Number 7 is simply another set of protocols that describe a means of communication between telephone switches in public telephone networks. They have been created and controlled by various bodies around the world, which leads to some specific local variations, but the principal organization with responsibility for their administration is the International Telecommunications Union or ITU-T.

Signaling System Number 7 (SS#7 or C7) is the protocol used by the telephone companies for interoffice signaling. In the past, in-band signaling techniques were used on interoffice trunks. This method of signaling used the same physical path for both the call-control signaling and the actual connected call. This method of signaling is inefficient and is rapidly being replaced by out-of-band or common-channel signaling techniques.

To understand SS7 we must first understand something of the basic inefficiency of previous signaling methods utilized in the Public Switched Telephone Network (PSTN).

A network utilizing common-channel signaling is actually two networks in one:

1. First there is the circuit-switched "user" network which actually carries the user voice and data traffic. It provides a physical path between the source and destination.
2. The second is the signaling network which carries the call control traffic. It is a packet-switched network using a common channel switching protocol.

The original common channel interoffice signaling protocols were based on Signaling System Number 6 (SS#6). Today SS#7 is being used in new installations worldwide. SS#7 is the defined interoffice signaling protocol for ISDN. The primary function of SS#7 is to provide call control, remote network management, and maintenance capabilities for the inter- office telephone network. SS#7 performs these functions by exchanging control messages between SS#7 telephone exchanges (signaling points or SPs) and SS#7 Signaling Transfer Points (STPs).

The Switching Offices (SPs) handle the SS#7 control networks as well as the user circuit-switched network. Basically, the SS#7 control network tells the switching office which paths to establish over the circuit-switched network. The STPs route SS#7 control packets across the signaling network. A switching office may or may not be an STP.

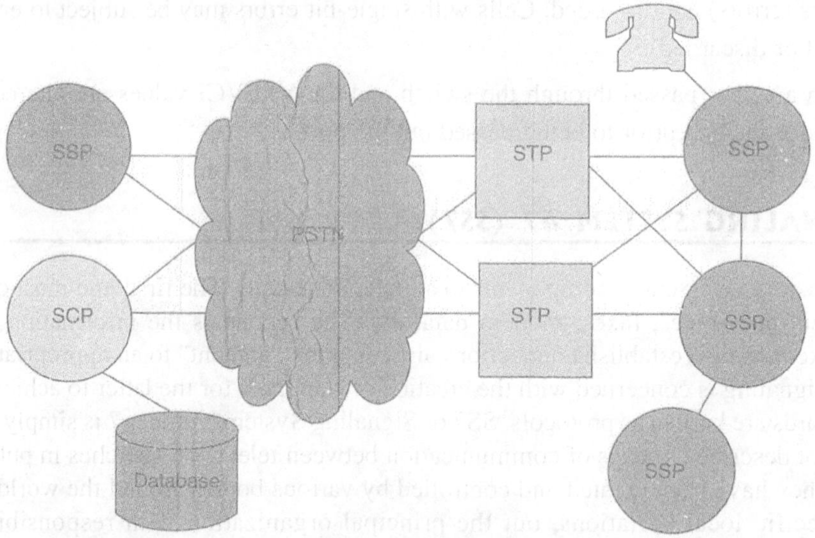

SSP : Service Switching Point
SCP : Service Control Point
STP : Service Transfer Point

Typical SS7 Architecture
or
Typical G7 Architecture

Fig. 4.7

The SS7 network and protocol are used for:

- Basic call setup, management, and tear down
- Wireless services such as Personal Communications Services (PCS), wireless roaming, and mobile subscriber authentication
- Local Number Portability (LNP)
- Toll-free (800/888) and toll (900) wire line services
- Enhanced call features such as call forwarding, calling party name/number display, and three-way calling
- Efficient and secure worldwide telecommunications
- Learn about our SS7/IP signaling products.

4.8.1 SS7 Protocol Layers

The SS7 network is an interconnected set of network elements that is used to exchange messages in support of telecommunications functions. The SS7 protocol is designed to both facilitate these functions and to maintain the network over which they are provided. Like most modern protocols, the SS7 protocol is layered.

Physical Layer (MTP-1)

This defines the physical and electrical characteristics of the signaling links of the SS7 network. Signaling links utilize DS-0 channels and carry raw signaling data at a rate of 56 kbps or 64 kbps (56 kbps is the more common implementation).

Message Transfer Part Level 2 (MTP-2)

The level 2 portion of the message transfer part (MTP Level 2) provides link-layer functionality. It ensures that the two end points of a signaling link can reliably exchange signaling messages. It incorporates such capabilities as error checking, flow control, and sequence checking.

Message Transfer Part Level 3 (MTP-3)

The level 3 portion of the message transfer part (MTP Level 3) extends the functionality provided by MTP level 2 to provide network layer functionality. It ensures that messages can be delivered between signaling points across the SS7 network regardless of whether they are directly connected. It includes such capabilities as node addressing, routing, alternate routing, and congestion control.

Signaling Connection Control Part (SCCP)

The Signaling Connection Control Part (SCCP) provides two major functions that are lacking in the MTP. The first of these is the capability to address applications within a signaling point. The MTP can only receive and deliver messages from a node as a whole; it does not deal with software applications within a node.

While MTP network-management messages and basic call-setup messages are addressed to a node as a whole, other messages are used by separate applications (referred to as subsystems) within a node. Examples of subsystems are 800 call processing, calling-card processing, Advanced Intelligent Network (AIN), and Custom Local-Area Signaling Services (CLASS) services (*e.g.*, repeat dialing and call return). The SCCP allows these subsystems to be addressed explicitly.

ISDN User Part (ISUP)

ISUP user part defines the messages and protocol used in the establishment and tear down of voice and data calls over the Public Switched Network (PSN), and to manage the trunk network on which they rely. Despite its name, ISUP is used for both ISDN and non-ISDN calls. In the North American version of SS7, ISUP messages rely exclusively on MTP to transport messages between concerned nodes.

Transaction Capabilities Application Part (TCAP)

TCAP defines the messages and protocol used to communicate between applications (deployed as subsystems) in nodes. It is used for database services such as calling card, 800, and AIN as well as switch-to-switch services including repeat dialing and call return. Because TCAP messages must be delivered to individual applications within the nodes they address, they use the SCCP for transport.

Operations, Maintenance and Administration Part (OMAP)

OMAP defines messages and protocol designed to assist administrators of the SS7 network. To date, the most fully developed and deployed of these capabilities are procedures for validating network routing tables and for diagnosing link troubles. OMAP includes messages that use both the MTP and SCCP for routing.

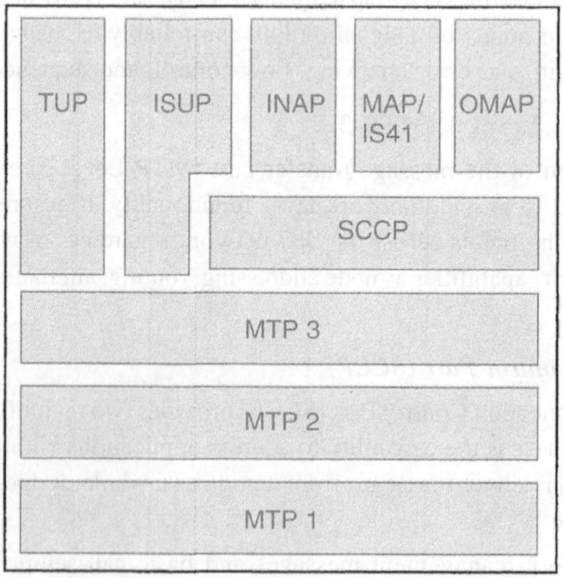

Fig. 4.8

4.8.2 Signaling Links

SS7 messages are exchanged between network elements over 56 or 64 kilobit per second (kbps) bidirectional channels called **signaling links.** Signaling occurs **out-of-band** on dedicated channels rather than **in-band** on voice channels. Compared to in-band signaling, out-of-band signaling provides:

- faster call setup times (compared to in-band signaling using Multi-Frequency (MF) signaling tones)
- more efficient use of voice circuits
- support for Intelligent Network (IN) services which require signaling to network elements without voice trunks (*e.g.,* database systems)
- improved control over fraudulent network usage
- Save network operating costs by reducing SS7 links.

4.8.3 Signaling Points

Each signaling point in the SS7 network is uniquely identified by a numeric **point code.** Point codes are carried in signaling messages exchanged between signaling points to identify the source and destination of each message. Each signaling point uses a routing table to select the appropriate signaling path for each message.

There are three kinds of signaling points in the SS7 network (Fig. 4.9):

- **SSP** (Service Switching Point)
- **STP** (Signal Transfer Point)
- **SCP** (Service Control Point)

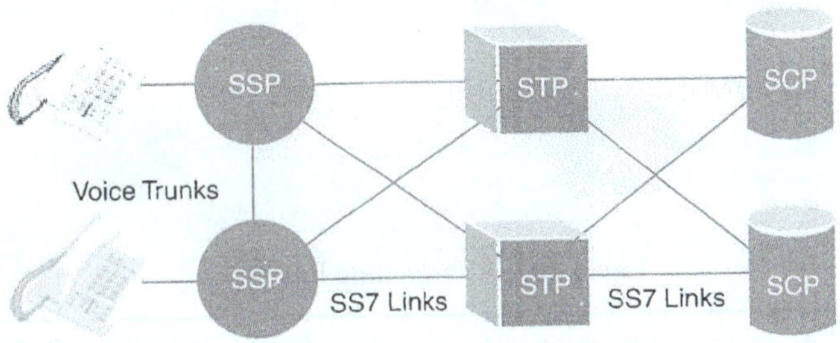

Fig. 4.9. SS7 Signaling points

SSPs are switches that originate, terminate, or tandem calls. An SSP sends signaling messages to other SSPs to setup, manage, and release voice circuits required to complete a call. An SSP may also send a query message to a centralized database (an **SCP**) to determine how to route a call (*e.g.,* a toll-free 1-800/888 call in North America). An SCP sends a response to the originating SSP containing the routing number(s) associated with the dialed number. An alternate routing number may be used by the SSP if the primary number is busy or the call is unanswered within a specified time. Actual call features vary from network to network and from service to service.

Network traffic between signaling points may be routed via a packet switch called an **STP.** An STP routes each incoming message to an outgoing signaling link based on routing information contained in the SS7 message. Because it acts as a network hub, an STP provides improved utilization of the SS7 network by eliminating the need for direct links between signaling points. An STP may perform **global title translation,** a procedure by which the destination signaling point is determined from digits present in the signaling message (*e.g.,* the dialed 800 number, calling card number, or mobile subscriber identification number). An STP can also act as a "firewall" to screen SS7 messages exchanged with other networks.

Because the SS7 network is critical to call processing, SCPs and STPs are usually deployed in mated pair configurations in separate physical locations to ensure network-wide service in the event of an isolated failure. Links between signaling points are also provisioned in pairs. Traffic is shared across all links in the linkset. If one of the links fails, the signaling traffic is rerouted over another link in the **linkset.** The SS7 protocol provides both error correction and retransmission capabilities to allow continued service in the event of signaling point or link failures.

- Signaling gateways can be configured as an STP or SEP (Signaling End Point).

4.8.4 SS7 Signaling Link Types

Signaling links are logically organized by link type ("A" through "F") according to their use in the SS7 signaling network.

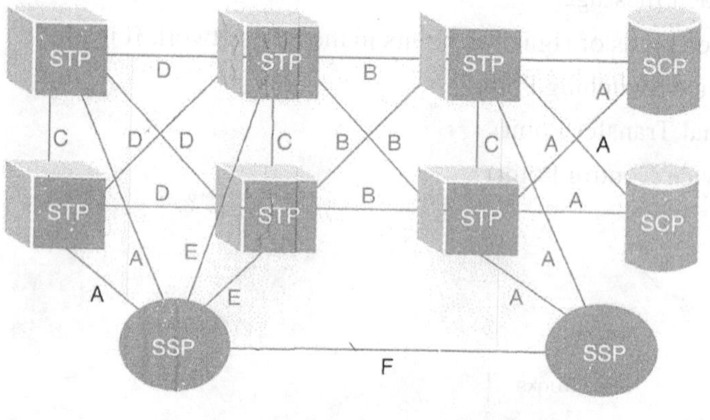

Fig. 4.10. SS7 signaling link types

A Link	An "A" (access) link connects a signaling end point (e.g., an SCP or SSP) to an STP. Only messages originating from or destined to the signaling end point are transmitted on an "A" link.
B Link	A "B" (bridge) link connects an STP to another STP. Typically, a quad of "B" links interconnect peer (or primary) STPs (*e.g.,* the STPs from one network to the STPs of another network). The distinction between a "B" link and a "D" link is rather arbitrary. For this reason, such links may be referred to as "B/D" links.

C Link	A "C" (cross) link connects STPs performing identical functions into a mated pair. A "C" link is used only when an STP has no other route available to a destination signaling point due to link failure(s). Note that SCPs may also be deployed in pairs to improve reliability; unlike STPs, however, mated SCPs are not interconnected by signaling links.
D Link	A "D" (diagonal) link connects a secondary (e.g., local or regional) STP pair to a primary (e.g., inter-network gateway) STP pair in a quad-link configuration. Secondary STPs within the same network are connected via a quad of "D" links. The distinction between a "B" link and a "D" link is rather arbitrary. For this reason, such links may be referred to as "B/D" links.
E Link	An "E" (extended) link connects an SSP to an alternate STP. "E" links provide an alternate signaling path if an SSP's "home" STP cannot be reached via an "A" link. "E" links are not usually provisioned unless the benefit of a marginally higher degree of reliability justifies the added expense.
F Link	An "F" (fully associated) link connects two signaling end points (*i.e.*, SSPs and SCPs). "F" links are not usually used in networks with STPs. In networks without STPs, "F" links directly connect signaling points.

4.8.5 SS7 Protocol Stack

The hardware and software functions of the SS7 protocol are divided into functional abstractions called "levels". These levels map loosely to the Open Systems Interconnect (OSI) 7-layer model defined by the International Standards Organization (ISO).

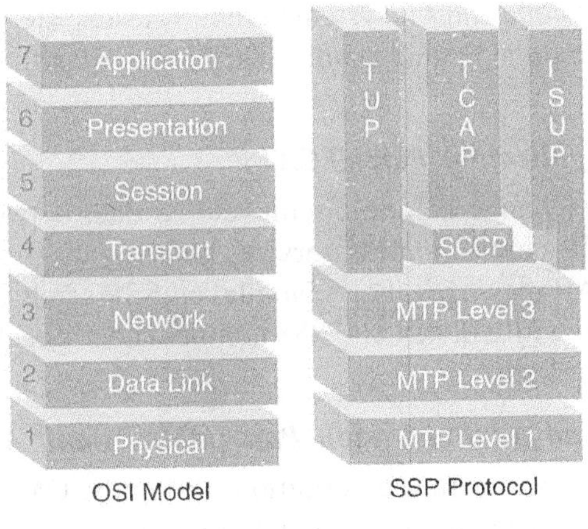

Fig. 4.11. The OSI reference model and the SS7 protocol stack

Message Transfer Part

The Message Transfer Part (MTP) is divided into three levels. The lowest level, MTP Level 1, is equivalent to the OSI Physical Layer. MTP Level 1 defines the physical, electrical, and functional characteristics of the digital signaling link. Physical interfaces defined include **E-1** (2048 kb/s; 32 64 kb/s channels), **DS-1** (1544 kb/s; 24 64kb/s channels), **V.35** (64 kb/s), **DS-0** (64 kb/s), and **DS-0A** (56 kb/s).

MTP Level 2 ensures accurate end-to-end transmission of a message across a signaling link. Level 2 implements flow control, message sequence validation, and error checking. When an error occurs on a signaling link, the message (or set of messages) is retransmitted. MTP Level 2 is equivalent to the OSI Data Link Layer.

MTP Level 3 provides message routing between signaling points in the SS7 network. MTP Level 3 re-routes traffic away from failed links and signaling points and controls traffic when congestion occurs. MTP Level 3 is equivalent to the OSI Network Layer.

ISDN User Part (ISUP)

The ISDN User Part (ISUP) defines the protocol used to set-up, manage, and release trunk circuits that carry voice and data between terminating line exchanges (*e.g.*, between a calling party and a called party). ISUP is used for both ISDN and non-ISDN calls. However, calls that originate and terminate at the same switch do not use ISUP signaling.

Telephone User Part (TUP)

In some parts of the world (*e.g.*, China, Brazil), the Telephone User Part (TUP) is used to support basic call setup and tear-down. TUP handles analog circuits only. In many countries, ISUP has replaced TUP for call management.

Signaling Connection Control Part (SCCP)

SCCP provides connectionless and connection-oriented network services and Global Title Translation (GTT) capabilities above MTP Level 3. A global title is an address (*e.g.*, a dialed 800 numbers, calling card number, or mobile subscriber identification number) which is translated by SCCP into a destination point code and subsystem number. A subsystem number uniquely identifies an application at the destination signaling point. SCCP is used as the transport layer for TCAP-based services.

Transaction Capabilities Application Part (TCAP)

TCAP supports the exchange of non-circuit related data between applications across the SS7 network using the SCCP connectionless service. Queries and responses sent between SSPs and SCPs are carried in TCAP messages. For example, an SSP sends a TCAP query to determine the routing number associated with a dialed 800/888 number and to check the Personal Identification number (PIN) of a calling card user.

Operations, Maintenance and Administration Part (OMAP) and ASE

OMAP and ASE are areas for future definition. Presently, OMAP services may be used to verify network routing databases and to diagnose link problems.

UNIT 5 MOBILE IP AND WIRELESS APPLICATION PROTOCOL

5.1 INTRODUCTION MOBILE IP

Mobile IP (or **IP mobility**) is an Internet Engineering Task Force (IETF) standard communications protocol that is designed to allow mobile device users to move from one network to another while maintaining a permanent IP address.The Mobile IP protocol allows location-independent routing of IP datagrams on the Internet. Each mobile node is identified by its home address disregarding its current location in the Internet. While away from its home network, a mobile node is associated with a *care-of* address which identifies its current location and its home address is associated with the local endpoint of a tunnel to its *home agent.* Mobile IP specifies how a mobile node registers with its home agent and how the home agent routes datagrams to the mobile node through the tunnel.

Mobile IP provides an efficient, scalable mechanism for roaming within the Internet. Using Mobile IP, nodes may change their point-of-attachment to the Internet without changing their home IP address. This allows them to maintain transport and higher-layer connections while roaming. Node mobility is realized without the need to propagate host-specific routes throughout the Internet routing fabric.

5.1.1 How Mobile IP Works?

A mobile node can have two addresses—a permanent home address and a care of address (CoA), which is associated with the network the mobile node is visiting. There are two kinds of entities in Mobile IP:

- A *home agent* stores information about moLile nodes whose permanent home address is in the home agent's network.
- A *foreign agent* stores information about mobile nodes visiting its network. Foreign agents also advertise care-of addresses, which are used by Mobile IP.

A node wanting to communicate with the mobile node uses the permanent home address of the mobile node as the destination address to send packets to. Because the home address logically belongs to the network associated with the home agent, normal IP routing mechanisms forward these packets to the home agent. Instead of forwarding these packets to a destination that is physically in the same network as the home agent, the home agent redirects these packets towards the foreign agent through an IP tunnel by encapsulating the datagram with a new IP header using the care of address of the mobile node.

When acting as transmitter, a mobile node sends packets directly to the other communicating node through the foreign agent, without sending the packets through the home agent, using its permanent home address as the source address for the IP packets. This is known as triangular routing. If needed, the foreign agent could employ *reverse tunneling* by tunneling the mobile node's packets to the home agent, which in turn forwards them to the communicating node. This is needed in networks whose gateway routers have ingress filtering enabled and hence the source IP address of the mobile host would need to belong to the subnet of the foreign network or else the packets will be discarded by the router.

The Mobile IP protocol defines the following:

- an authenticated registration procedure by which a mobile node informs its home agent(s) of its care-of-address"(es);
- an extension to ICMP Router Discovery, which allows mobile nodes to discover prospective home agents and foreign agents; and
- the rules for routing packets to and from mobile nodes, including the specification of one mandatory tunneling mechanism and several optional tunneling mechanisms.

Operation of Mobile IP

Mobil node is assigned to a particular network - home network

- IP address on home network is static - home address
- Mobile node can move to another network - foreign network
- Mobile node registers with network node on foreign network - foreign agent
- Mobile node gives care-of address to agent on home network - home agent.

5.2 APPLICATIONS

Mobile IP is most often found in wired and wireless environments where users need to carry their mobile devices across multiple LAN subnets. It may for example be used in roaming between overlapping wireless systems, for example IP over DVB, WLAN, WiMAX and BWA. Currently, Mobile IP is not required within cellular systems such as 3G, to provide transparency when Internet users migrate between cellular towers, since these systems provide their own data link layer handover and roaming mechanisms. However, it is often used in 3G systems to allow seamless IP mobility between different Packet Data Serving Node (PDSN) domains.

In many applications (*e.g.*, VPN, VoIP), sudden changes in network connectivity and IP address can cause problems.

5.2.1 Mobile IP Uses

- Enable computers to maintain Internet connectivity while moving from one Internet attachment point to another
- Mobile - user's point of attachment changes dynamically and all connections are automatically maintained despite the change
- Nomadic - user's Internet connection is terminated each time the user moves and a new connection is initiated when the user dials back in new, temporary IP address is assigned.

5.2.2 Capabilities of Mobile IP

- Discovery - mobile node uses discovery procedure to identify prospective home and foreign agents
- Registration - mobile node uses an authenticated registration procedure to inform Home agent of its care-of address
- Tunneling - used to forward IP datagrams from a home address to a care-of address.

Move Detection

Mobile node may move from one network to another due to some handoff mechanism without IP level being aware:

- Agent discovery process is intended to enable the agent to detect such a move
- Algorithms to detect move
- Use of lifetime field - mobile node uses lifetime field as a timer for agent advertisements
- Use of network prefix - mobile node checks if any newly received agent advertisement messages are on the same network as the node's current care-of address.

Co-Located Addresses

If mobile node moves to a network that has no foreign agents, or all foreign agents are busy, it can act as its own foreign agent:

- Mobile agent uses co-located care-of address
- IP address obtained by mobile node associated with mobile node's current network interface
- Means to acquire co-located address
- Temporary IP address through an Internet service, such as DHCP
- May be owned by the mobile node as a long-term address for use while visiting a given foreign network.

Registration Process

- Mobile node sends registration request to foreign agent requesting forwarding service
- Foreign agent relays request to home agent
- Home agent accepts or denies request and sends registration reply to foreign agent
- Foreign agent relays reply to mobile node.

Registration Operation Messages

- Registration request message
  - o Fields = type, S, B, D, M, V, G, lifetime, home, address, home agent, care-of-address, identification, extensions
- Registration reply message
  - o Fields = type, code, lifetime, home address, home, agent, identification, extensions.

Registration Procedure Security

Mobile IP designed to resist attacks

- Node pretending to be a foreign agent sends registration request to a home agent to divert mobile node traffic to itself
- Agent replays old registration messages to cut mobile node from network for message authentication, registration request and reply contain authentication extension
- Fields = type, length, Security Parameter Index (SPI), authenticator.

Types of Authentication Extensions

- Mobile-home - provides for authentication of registration messages between mobile node and home agent; must be present
- Mobile-foreign - may be present when a security association exists between mobile node and foreign agent
- Foreign-home - may be present when a security association exists between foreign agent and home agent.

Tunneling

- Home agent intercepts IP datagrams sent to mobile node's home address
  - o Home agent informs other nodes on home network that datagrams to mobile node should be delivered to home agent
- Datagrams forwarded to care-of address via tunneling
  - o Datagram encapsulated in outer IP datagram.

Mobile IP Encapsulation Options

- IP-within-IP—entire IP datagram becomes payload in new IP datagram
  - o Original, inner IP header unchanged except TTL decremented by 1
  - o Outer header is a full IP header
- Minimal encapsulation—new header is inserted between original IP header and original IP payload
  - o Original IP header modified to form new outer IP header
- Generic Routing Encapsulation (GRE) - developed prior to development of Mobile IP.

5.3 WIRELESS APPLICATION PROTOCOL

5.3.1 An Overview of Wireless Application Protocol

The Wireless Application Protocol (WAP) is a hot topic that has been widely hyped in the mobile industry and outside of it. WAP is simply a protocol—a standardized way that a mobile phone talks to a server installed in the mobile phone network.

The Wireless Application Protocol (WAP) is an important development in the wireless industry because of its attempt to develop an open standard for wireless protocols, independent of vendor and airlink.

WAP is Hot for Several Reasons

The Wireless Application Protocol (WAP) is a hot topic that has been widely hyped in the mobile industry and outside of it. WAP is simply a protocol- a standardized way that a mobile phone talks to a server installed in the mobile phone network.

WAP Services

- It provides a standardized way of linking the Internet to mobile phones, thereby linking two of the fastest growing industries anywhere in the world.
- Its founder members include the major wireless vendors of Nokia, Ericsson and Motorola, plus a newcomer Phone.com
- The WAP Forum has over 120 member companies
- Mobile information services, a key application for WAP, have not been as successful as many network operators expected. WAP is seen as a way to rectify this situation.

5.3.2 WAP Formation and Philosophy

Formation

Motorola, Nokia, Ericsson and the US software company Phone.com (formerly Unwired Planet) were the initial partners that teamed up over two years ago in mid 1997 to develop and deploy the Wireless Application Protocol (WAP). WAP is an attempt to define the standard for how content from the Internet is filtered for mobile communications. Content is now readily available on the Internet and WAP was designed as the (rather than one) way of making it easily available on mobile terminals.

The WAP Forum was formed after a US network operator Omni point issued a tender for the supply of mobile information services in early 1997. It received several responses from different suppliers using proprietary techniques for delivering the information such as Smart Messaging from Nokia and HDML from Phone.com (then called Unwired Planet). Omni point informed the tender responders that it would not accept a proprietary approach and recommended that various vendors get together to explore defining a common standard. After all, there was not a great deal of difference between the different approaches, which could be combined and extended to form a powerful standard. These events were the initial stimulus behind the development of the Wireless Application

Protocol, with Ericsson and Motorola joining Nokia and Unwired Planet as the founder members of the WAP Forum.

Philosophy

The Wireless Application Protocol takes a client server approach. It incorporates a relatively simple microbrowser into the mobile phone, requiring only limited resources on the mobile phone. This makes WAP suitable for thin clients and early smart phones. WAP puts the intelligence in the WAP Gateways whilst adding just a microbrowser to the mobile phones themselves. Microbrowser-based services and applications reside temporarily on servers, not permanently in phones. The Wireless Application Protocol is aimed at turning a mass-market mobile phone into a "network-based smart phone". The Wireless Application Protocol is envisaged as a comprehensive and scaleable protocol designed for use with:

- Any mobile phone from those with a one line display to a smart phone.
- Any existing or planned wireless service such as the Short Message Service, Circuit Switched Data, Unstructured Supplementary Services Data (USSD) and GPRS.
- The importance of WAP can be found in the fact that it provides an evolutionary path for application developers and network operators to offer their services on different network types, bearers and terminal capabilities. The design of the WAP standard separates the application elements from the bearer being used. This helps in the migration of some applications from SMS or Circuit Switched Data to GPRS for example.
- Any mobile network standard such as Code Division Multiple Access (CDMA), Global System for Mobiles (GSM), or Universal Mobile Telephone System (UMTS). WAP has been designed to work with all cellular standards and is supported by major worldwide wireless leaders such as AT and T Wireless and NTT DoCoMo.
- Multiple input terminals such as keypads, keyboards, touch-screens and styluses.

5.4 THE TYPES OF WIRELESS APPLICATION PROTOCOL

The Wireless Application Protocol embraces and extends the previously conceived and developed wireless data protocols. Phone.com created a version of the standard HTML (HyperText Markup Language) Internet protocols designed specifically for effective and cost-effective information transfer across mobile networks. Wireless terminals incorporated a HDML (Handheld Device Markup Language) microbrowser and Phone.com's Handheld Device Transport Protocol (HDTP) then linked the terminal to the UP. Link Server Suite which connected to the Internet or intranet where the information being requested resides. The Internet site content was tagged with HDML.

This technology was incorporated into WAP and renamed using some of the many WAP-related acronyms such as WMLS, WTP and WSP. Someone with a WAP-compliant phone uses the in-built microbrowser to:

1. Make a request in WML (Wireless Markup Language), a language derived from HTML especially for wireless network characteristics.

2. This request is passed to a WAP Gateway that then retrieves the information from an Internet server either in standard HTML format or preferably directly prepared for wireless terminals using WML. If the content being retrieved is in HTML format, a filter in the WAP Gateway may try to translate it into WML. A WML scripting language is available to format data such as calendar entries and electronic business cards for direct incorporation into the client device.
3. The requested information is then sent from the WAP Gateway to the WAP client, using whatever mobile network bearer service is available and most appropriate.

WAP Architecture

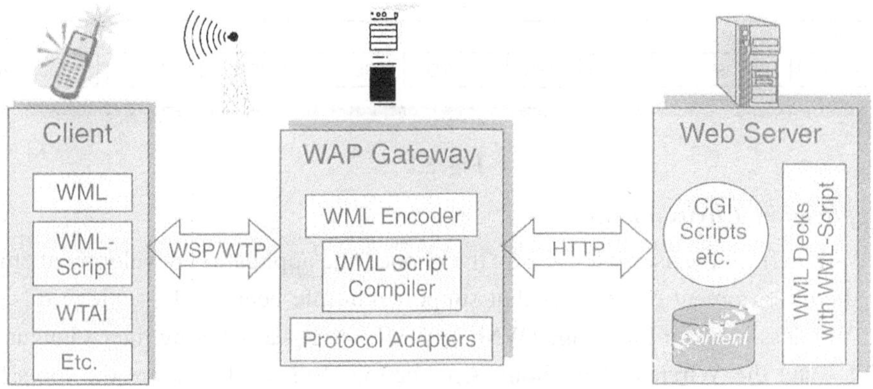

Fig. 5.1

WAP Protocol Stack Architecture

WAP has a layered architecture as shown in the diagram below:

Wireless Application Environment (WAE)
Wireless Session Protocol (WSP)
Wireless Transaction Protocol (WTP)
Wireless Transport Layer Security (WTLS)
Wireless Datagram Protocol (WDP)
Bearers e.g. Data, SMS, USSD

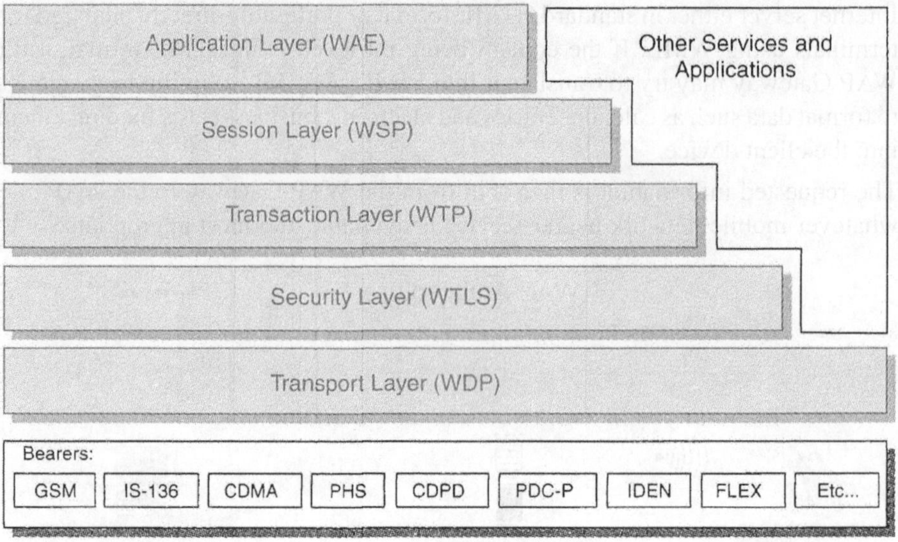

Fig. 5.2

Wireless Application Environment

The WAE defines the user interface on the phone. The application development environment to facilitate the development of services that support multiple bearers. To achieve this, the WAE contains the Wireless Markup Language (WML), WMLScript - a scripting micro-language similar to JavaScript - and the Wireless Telephony Application (WTA). These are the tools that allow WAP-based applications to be developed.

Wireless Session Protocol

A sandwich layer that links the WAE to two session services— one connection oriented operating above the Wireless Transaction Protocol and a connectionless service operating above the Wireless Datagram Protocol.

Wireless Transaction Protocol

Runs on top of a datagram service such as User Datagram Protocol (UDP); part of the standard suite of TCP/IP protocols, to provide a simplified protocol suitable for low bandwidth mobile stations. WTP offers three classes of transaction service: unreliable one way request, reliable one way request and reliable two way request respond. Interestingly, WTP supports Protocol Data Unit concatenation and delayed acknowledgement to help reduce the number of messages sent. This protocol therefore tries to optimize the user experience by providing the information that is needed when it is needed—it can be confusing to received confirmation of delivery messages when you are expecting the information itself. By stringing several messages together, the end user may well be able to get a better feel more quickly for what information is being communicated.

Wireless Transport Layer Security

WTLS incorporates security features that are based upon the established Transport Layer Security (TLS) protocol standard. Includes data integrity checks, privacy on the WAP Gateway to client leg and authentication.

Wirelesses Datagram Protocol

Allows WAP to be bearer independent by adapting the transport layer of the underlying bearer. WDP presents a consistent data format to the higher layers of the WAP protocol stack thereby conferring the advantage of bearer independence to application developers.

5.5 OPTIMAL WAP BEARER

- Short Message Service
- Circuit Switched Data
- Unstructured Supplementary Services Data
- General Packet Radio Service

Short Message Service

Given its limited length of 160 characters per short message, SMS may not be an adequate bearer for WAP because of the weight protocol of the protocol. The overhead of the WAP protocol that would be required to be transmitted in an SMS message would mean that even for the simplest of transactions several SMS messages may in fact have to be sent. This means that using SMS as a bearer can be a time consuming and expensive exercise. Only one network operator—SBC of the US—is known to be developing WAP services based on SMS.

Circuit Switched Data

Most of the trial WAP based services use CSD as the underlying bearer. Since CSD has relatively few users currently, WAP could kickstart usage of and traffic generated by this bearer.

However, CSD lacks immediacy—a dial up connection taking about 10 seconds is required to connect the WAP client to the WAP Gateway, and this is the best case scenario when there is an complete end to end digital call—in the case of the need for analog modem handshaking (because the WAP phone does not support V.110 the digital protocol, or the WAP Gateway does not have a digital direct connection such as ISDN into the mobile network), the connect time is increased to about 30 seconds.

Unstructured Supplementary Services Data

Unstructured Supplementary Services Data (USSD) is a means of transmitting information or instructions over a GSM network. USSD has some similarities with SMS since both use the GSM network's signaling path. Unlike SMS, USSD is not a store and forward service and is session-oriented such that when a user accesses a USSD service, a session is established and the radio connection stays open until the user, application, or time out releases it. This has more in common with Circuit Switched Data than SMS. USSD text messages can be up to 182 characters in length.

USSD has some advantages and disadvantages as a tool for deploying services on mobile networks:

Turnaround response times for interactive applications are shorter for USSD than SMS because of the session-based feature of USSD, and because it is NOT a store and forward service. According to Nokia, USSD can be up to seven times faster than SMS to carry out the same two-way transaction.

- Users do not need to access any particular phone menu to access services with USSD—they can enter the Unstructured Supplementary Services Data (USSD) command direct from the initial mobile phone screen.
- Because USSD commands are routed back to the home mobile network's Home Location Register (HLR), services based on USSD work just as well and in exactly the same way when users are roaming.
- Unstructured Supplementary Services Data (USSD) works on all existing GSM mobile phones.
- Both SIM Application Toolkit and the Wireless Application Protocol support USSD.
- USSD Stage 2 has been incorporated into the GSM standard. Whereas USSD was previously a one way bearer useful for administrative purposes such as service access, Stage 2 is more advanced and interactive. By sending in a USSD2 command, the user can receive an information services menu. As such, USSD Stage 2 provides WAP-like features on EXISTING phones.
- USSD strings are typically complicated for the user to remember, involving the use of the "*" and "#" characters to denote the start and finish of the USSD string. However, (USSD) strings for regularly used services can be stored in the phonebook, reducing the need to remember and re-enter them.

General Packet Radio Service

The General Packet Radio Service (GPRS) is a new packet-based bearer that is being introduced on many GSM and TDMA mobile networks from the year 2000 onwards. It is an exciting new bearer because it is immediate (there is no dial up connection), relatively fast (up to 177.2 kbps in the very best theoretical extreme) and supports virtual connectivity, allowing relevant information to be sent from the network as and when it is generated.

At the time of writing in early August 1999, there has been no confirmation from any handset vendors that mobile terminated GPRS traffic (*i.e.*, direct receipt of GPRS packets on the mobile phone) will be supported by the initial GPRS terminals. Availability or not of GPRS MT is a central question with critical impact on the GPRS business case such as application migration from other nonvoice bearers.

There are two efficient means of delivering proactively sending (pushing) content to a mobile phone: by the Short Message Service which is of course one of WAP bearers or by the user maintaining more or less a permanent GPRS (mobile originated) session with the content server. However, mobile terminated IP traffic might allow unsolicited information to reach the terminal. Internet sources originating such unsolicited content may not be chargeable. A possible worse case scenario would be that mobile users would have to pay for receiving unsolicited junk content. This is a potential reason for a mobile vendor NOT to support GPRS Mobile Terminate in their GPRS terminals. However, by originating the session themselves from their handset, users confirm their agreement to pay for the delivery of content from that service. The early WAP enabled phones will also support the General Packet Radio Service, WAP and GPRS could well be synergistic and be used widely together. For the kinds of interactive, menu based information exchanges that WAP anticipates, Circuit Switched Data is not immediate enough because of the need to set up a call. Early prototypes of WAP services based on Circuit Switched Data were therefore close to unusable. SMS on the other hand is immediate but is ALWAYS store and forward, such that even when a subscriber has just requested information from their microbrowser, the SMS Centre resources are used in the information transfer. As such, GPRS and WAP are ideal bearers for each other.

Additionally, WAP incorporates two different connection modes—**WSP connection mode or WSP connectionless protocol.** This is very similar to the two GPRS Point to Point services—connection oriented and connection less.

The predominant bearer for WAP-based services will depend on delays in availability of WAP handsets and delays in the availability of GPRS terminals. If WAP terminals are delayed until the year 2000, most WAP terminals will support GPRS as well. If the first WAP terminals support SMS and Circuit Switched Data, but not GPRS, then SMS could become the predominant initial WAP bearer.

WAP certainly will be important for the development of GPRS-based applications. Because the bearer level is separated from the application layer in the WAP protocol stack, WAP provides the ideal and defined and standardized means to port the same application to different bearers. As such, many application developers will use WAP to facilitate the migration of their applications across bearers once GPRS based WAP protocols are supported.

WAP Forum Members List

The initial Wireless Application Protocol partner companies - Nokia, Ericsson, Motorola and Phone.com (formerly Unwired Planet) - formed a limited company called WAP Forum Limited to administer the global Wireless Application Protocol specification process and get new companies involved in developing the protocol.

By August 1999, the WAP Forum had over 120 members comprising major phone manufacturers, network operators, SMS Centre suppliers and SMS software suppliers. Announced WAP Forum members include:

- Telecommunications Hardware
- Telecommunications Software
- Mobile Telephone Network Operators
- Smart Cards and Security

Applications

WAP is being used to develop enhanced forms of existing applications and new versions of today's applications. Existing mobile data software and hardware supplies are adding WAP support to their offering, either by developing their own WAP interface or more usually partnering with one of the WAP Gateway suppliers profiled above. WAP is also given a significant impetus for new players to add mobile as a new distribution channel for their existing products and services - for example, CNN and Nokia teamed up to offer CNN Mobile and Reuters and Ericsson teamed up to provide Reuters Wireless Services. The Wireless Application Protocol will allow customers to easily reply to incoming information on the phone by allowing new menus to access mobile services. This is part of the business case for network operators - by making the value-added services more easily to reply to and request (using menus instead of keywords, for example), WAP can help generate additional traffic on the network and therefore revenue.

WAP facilitates easy migration of applications between networks and bearers. As such, WAP is similar to Java in that it simplifies application development. This reduces the cost of wireless application development and therefore encourages entry to the mobile industry by software developers.

- Corporate Applications
- Consumer Applications

Corporate Applications

- Corporate applications that are being enhanced and enabled with a WAP interface include:
- Job Dispatch
- Remote Point Of Sale
- Customer Service
- Remote Monitoring Such As Meter Reading
- Vehicle Positioning
- Corporate Email
- Remote LAN Access
- File Transfer
- Web Browsing
- Document Sharing/Collaborative Working
- Audio
- Still Images
- Moving Images
- Home Automation

Consumer Applications

- Consumer applications that are being enhanced and enabled with a WAP interface include:
- Simple Person to Person Messaging
- Voice and Fax Mail Notifications
- Unified Messaging
- Internet Email
- Prepayment
- Ring tones

- Mobile Commerce
- Affinity Programs
- Mobile Banking
- Chat
- Information Services.

5.6 WIRELESS DATAGRAM PROTOCOL

5.6.1 Introduction

The WDP offers a consistent service at the Transport Service Access Point to the upper layer protocol of WAP. This consistency of service allows for applications to operate transparently over different available bearer services. The varying heights of each of the bearer services and thus the difference in WDP protocol necessary to operate over those bearers to maintain the same service offering at the Transport Service Access Point is accomplished by a bearer adaptation. WDP can be mapped onto different bearers, with different characteristics.

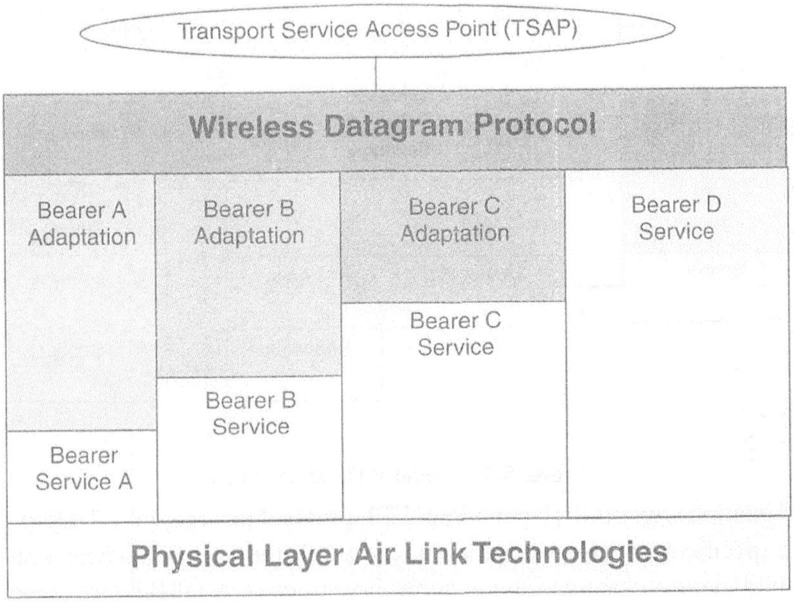

Fig. 5.3

Wireless Datagram Protocol Architecture

The services offered by WDP include application addressing by port numbers, optional segmentation and reassembly and optional error detection. The services allow for applications to operate transparently over different available bearer services.

In order to optimize the protocol with respect to memory usage and radio transmission efficiency, the protocol performance over each bearer may vary. However, the WDP service and service primitives will remain the same, providing a consistent interface to the higher layers.

The WDP layer operates above the data capable bearer services supported by the various network types. As a general datagram service, WDP offers a consistent service to the upper layer protocol (Security, Transaction and Session) of WAP and communicate transparently over one of the available bearer services.

WDP supports several simultaneous communication instances from a higher layer over a single underlying WDP bearer service. The port number identifies the higher layer entity above WDP. This may be another protocol layer such as the Wireless Transaction Protocol (WTP) or the Wireless Session Protocol (WSP) or an application such as electronic mail. By reusing the elements of the underlying bearers, WDP can be implemented to support multiple bearers and yet be optimized for efficient operation within the limited resources of a mobile device.

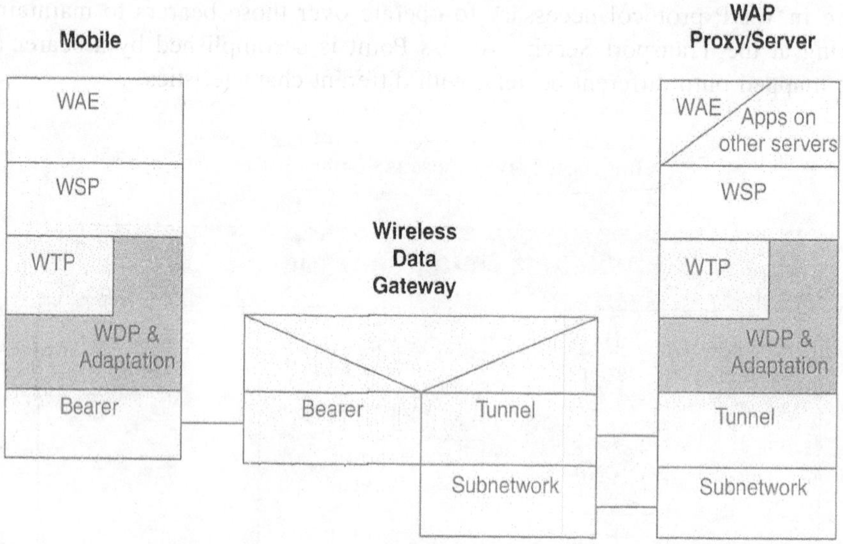

Fig. 5.4. General WDP architecture

The Adaptation Layer is the layer of the WDP protocol that maps the WDP protocol functions directly onto a specific bearer. The Adaptation Layer is different for each bearer and deals with the specific capabilities and characteristics of that bearer service. The Bearer Layer is the bearer service such as GSM SMS, or USSD, or IS- 136 R-Data, or CDMA Packet Data. At the Gateway the Adaptation Layer terminates and passes the WDP packets on to a WAP Proxy/Server via a Tunneling protocol, which is the interface between the Gateway that supports the bearer service and the WAP Proxy/Server. For example if the bearer were GSM SMS, the Gateway would be a GSM SMSC and would support a specific protocol (the Tunneling protocol) to interface the SMSC to other servers. The Sub Network is any common networking technology that can be used to connect two communicating devices, examples are wide-area networks based on TCP/IP or X.25, or LANs operating TCP/IP over Ethernet. The WAP Proxy/Server may offer application content or may act as a gateway between the wireless WTP protocol suites and the wired Internet.

5.6.2 WDP Management Entity

The WDP Management Entity is used as an interface between the WDP layer and the environment of the device. The WDP Management Entity provides information to the WDP layer about changes in the devices environment, which may impact the correct operation of WDP.

The WDP protocol is designed around an assumption that the operating environment is capable of transmitting and receiving data.

For example, this assumption includes the following basic capabilities that must be provided by the mobile:

- the mobile is within a coverage area applicable to the bearer service being invoked;
- the mobile having sufficient power and the power being on;
- sufficient resources (processing and memory) within the mobile are available to WDP;
- the WDP protocol is correctly configured, and;
- the user is willing to receive/transmit data.

5.6.3 Processing Errors of WDP Datagram's

Processing errors can happen when WDP datagram's are sent from a WDP provider to another. For example, a Wireless Data Gateway may not be able to send the datagram to the WAP Gateway, or there is no application listening to the destination port, or the receiver might not have enough buffer space to receive a large message.

The Wireless Control Message Protocol (WCMP) provides an efficient error handling mechanism for WDP, resulting in improved performance for WAP protocols and applications. Therefore the WCMP protocol SHOULD be implemented. See the [WCMP] specification.

5.7 WIRELESS TRANSACTION PROTOCOL

5.7.1 Protocol Features

The following list summarizes the features of WTP. Three classes of transaction service:

(*i*) Class 0: Unreliable invoke message with no result message

(*ii*) Class 1: Reliable invoke message with no result message

(*iii*) Class 2: Reliable invoke message with exactly one reliable result message

- Reliability is achieved through the use of unique transaction identifiers, acknowledgements, duplicate removal and re-transmissions.
- No explicit connections set up or tear down phases. Explicit connection open and/or close imposes excessive overhead on the communication link.

- Optionally user-to-user reliability: the WTP user confirms every received message.
- Optionally, the last acknowledgement of the transaction MAY contain out of band information related to the transaction. For example, performance measurements.
- Concatenation MAY be used, where applicable, to convey multiple Protocol Data Units in one Service Data Unit of the datagram transport.
- Message orientation. The basic unit of interchange is an entire message and not a stream of bytes.
- The protocol provides mechanisms to minimize the number of transactions being replayed as the result of duplicate packets.
- Abort of outstanding transaction, including flushing of unsent data both in client and server. The abort can be triggered by the user canceling a requested service.
- For reliable invoke messages, both success and failure is reported. If an invoke can not be handled by the Responder, an abort message will be returned to the Initiator instead of the result.
- The protocol allows for asynchronous transactions. The Responder sends back the result as the data becomes available.

5.7.2 Relation to Other Protocols

WTP is specified to run over a datagram transport service. The WTP protocol data unit is located in the data portion of the datagram. Since datagram's are unreliable, WTP is required to perform re-transmissions and send acknowledgement in order to provide a reliable service to the WTP user. The services provided to the WTP user are located.

WTP User (*e.g.*, WSP)	
WTP	❑ Transaction handling ❑ Re-transmissions, duplicate removal, acknowledgements ❑ Concatenation and separation
[WTLS]	❑ Optionally compression ❑ Optionally encryption ❑ Optionally authentication
Datagram Transport (*e.g.*, WDP)	❑ Port number addressing ❑ Segmenation and re-assembly (if provided) ❑ Error detection (if provided)
Bearer Network (*e.g.*, IP, GSM SMS/USSD, IS-136 GUTS)	❑ Routing ❑ Device addressing (IP address, MSISDN) ❑ Segmentation and re-assembly (if provided) ❑ Error detection (if provided)

WTP is also responsible for concatenation (if possible) of multiple protocol data units into one transport service data unit.

Security Considerations

WTP has no security mechanisms.

5.7.3 Management Entity

The WTP Management Entity is used as an interface between the WTP layer and the environment of the device. The WTP Management Entity provides information to the WTP layer about changes in the device environment, which MAY impact the correct operation of WTP.

The WTP protocol is designed around an assumption that the environment in which it is operating is capable of transmitting and receiving data. For example, this assumption includes the following basic capabilities that MUST be provided by the mobile device:

- the mobile is within a coverage area applicable to the bearer service being invoked;
- the mobile having sufficient power and the power being on;
- sufficient resources (processing and memory) within the mobile are available to WTP;
- the WTP protocol is correctly configured, and;
- The user is willing to receive/transmit data.

The WTP Management Entity monitors the state of the above services/capabilities of the mobile's environment and would notify the WTP layer if one or more of the assumed services were not available. For example if the mobile roamed out of coverage for a bearer service, the Bearer Management Entity SHOULD report to the WTP Management

Entity that transmission/reception over that bearer is no longer possible. In turn, the WTP Management Entity would indicate to the WTP layer to close all active connections over that bearer. Other examples such as low battery power would be handled in a similar way by the WTP Management Entity.

Other WTP Users

The intended use of this protocol is to provide WSP [WSP] with a reliable transaction service over an unreliable datagram service. However, the protocol can be used by other applications with similar communication needs.

5.8 WIRELESS SESSION PROTOCOL

5.8.1 WSP Architectural Overview

Wireless Session Protocol is a session-level protocol family for remote operations between a client and proxy or server. Reference Model.

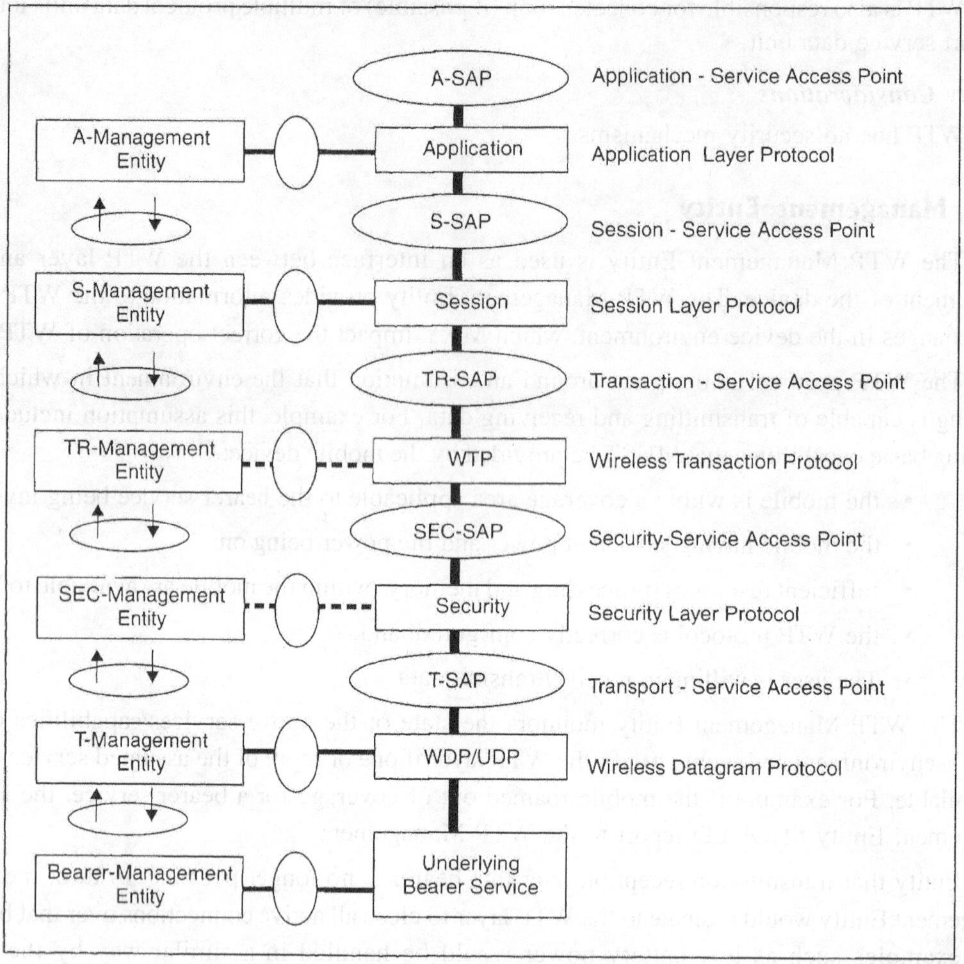

Fig. 5.5. Wireless application protocol reference model

A model of layering the protocols in WAP is illustrated in Fig. 5.5. WAP protocols and their functions are layered in a style resembling that of the ISO OSI Reference Model [ISO7498]. Layer Management Entities handle protocol initialization, configuration and error conditions (such as loss of connectivity due to the mobile station roaming out of Coverage) that are not handled by the protocol itself.

WSP is designed to function on the transaction and datagram services. Security is assumed to be an optional layer above the transport layer. The security layer preserves the transport service interfaces. The transaction, session or application management entities are assumed to provide the additional support that is required to establish security Contexts and secure connections. This support is not provided by the WSP protocols directly. In this regard, the security layer is modular. WSP itself does not require a security layer; however, applications that use WSP may require it.

WSP Features

WSP provides a means for organized exchange of content between co-operating client/server applications. Specifically, it provides the applications means to:

(*a*) Establish a reliable session from client to server and release that session in an orderly manner;

(*b*) Agree on a common level of protocol functionality using capability negotiation;

(*c*) Exchange content between client and server using compact encoding;

(*d*) Suspend and resume the session.

The currently defined services and protocols (WSP) are most suited for browsing-type applications. WSP defines actually two protocols: one provides connection-mode session services over a transaction service, and another provides non-confirmed, connectionless services over a datagram transport service. The connectionless service is most suitable.

When applications do not need reliable delivery of data and do not care about confirmation. It can be used without actually having to establish a session.

In addition to the general features, WSP offers means to:

(*a*) *Provide HTTP/1.1 Functionality:*

1. Extensible request-reply methods
2. Composite objects
3. Content type negotiation.

(*b*) Exchange client and server session headers

(*c*) Interrupt transactions in process

(*d*) Push content from server to client in an unsynchronized manner

(*e*) Negotiate support for multiple, simultaneous asynchronous transactions.

5.8.2 Basic Functionality

The core of the WSP design is a binary form of HTTP. Consequently the requests sent to a server and responses going to a client may include both headers (meta-information) and data. All the methods defined by HTTP/1.1 are supported.

In addition, capability negotiation can be used to agree on a set of extended request methods, so that full compatibility to HTTP/1.1 applications can be retained.WSP provides typed data transfer for the application layer. The HTTP/1.1 content headers are used to define content type, character set encoding, languages, etc, in an extensible manner. However, compact binary encodings are defined for the well-known headers to reduce protocol overhead. WSP also specifies a compact composite data format that provides content headers for each component within the composite data object. This is a semantically equivalent binary form of the MIME "multipart/mixed" format used by HTTP/1.1.

The lifecycle of a WSP session is not tied to the underlying transport. A session can be suspended while the session is idle to free up network resources or save battery. A lightweight session re-establishment protocol allows the session to be resumed without the overhead of full-blown session establishment. A session may be resumed over a different bearer network.

Extended Functionality

WSP allows extended capabilities to be negotiated between the peers. This allows for high-performance, feature full implementation as well as simple, basic and small implementations.WSP provides an optional mechanism for attaching header information (meta-data) to the acknowledgement of a transaction. This allows the client application to communicate specific information about the completed transaction back to the server.

WSP provides both push and pull data transfer. Pull is done using the request/response mechanism from HTTP/1.1. In addition,

WSP provides three push mechanisms for data transfer:

- Confirmed data push within an existing session context
- Non-confirmed data push within an existing session context
- Non-confirmed data push without an existing session
  - o The confirmed data push mechanism allows the server to push data to the client at any time during a session. The server receives confirmation that the push was delivered.
  - o The non-confirmed data push within an existing session provides a similar function as reliable data push, but without confirmation. The non-confirmed data push can also without an existing session. In this case, a default session context is assumed. Non-confirmed out-of-session data push can be used to send one-way messages over an unreliable transport.

WSP optionally supports asynchronous requests, so that a client can submit multiple requests to the server simultaneously. This improves utilization of airtime in that multiple requests and replies can be coalesced into fewer messages. This also improves latency as the results of each request can be sent to the client when it becomes available.

WSP partitions the space of well-known header field names into *header code pages*. Each code page can define only a fairly limited number of encodings for well-known field names, which permits them to be represented more compactly. Running out of identities for well-known field names on a certain code page is still not a problem, since WSP specifies a mechanism for shifting from one header code page to another.

5.8.3 WSP Protocol Operations

- Utilisation of WTP
- Protocol Description
- Protocol Parameters
- Variables.
- Event Processing
- State Tables

Connection-Mode WSP

This section describes the operations of WSP over the WTP transaction service [WTP].

Utilization of WTP

The WTP transaction classes utilized by each WSP facility are summarized in Table 1.

Table 5.1. Utilization of WTP

WSP Facility	*WTP Transaction Classes*
Session Management	Class 0 and Class 2
Method Invocation	Class 2
Session Resume	Class 0 and Class 2
Push	Class 0
Confirmed Push	Class 1

A connection-mode WSP client MUST support initiation of WTP Class 0 and Class 2 transactions. The client SHOULD accept Class 0 transaction invocations from the server, so that the server is able to disconnect the session explicitly. If the client is to support the push facilities, it MUST accept transactions in the class, which the table above defines to be used by each push facility.

Protocol Description

The following diagrams illustrate the use of a transaction service by the session facilities. The specific details of how the protocol works are expressed in the state tables in section 5.8-2.1, "State Tables", below. Any discrepancy between the diagrams and the state tables shall be decided in favour of the state tables. The dashed arrows represent the WTP protocol messages carrying acknowledgements and WSP PDUs as their data; the messages indicated by parallel arrows are likely to be concatenated into a single transport datagram.

Session Management Facility

Normal session creation proceeds without any error or redirection as shown in Fig. 5.6.

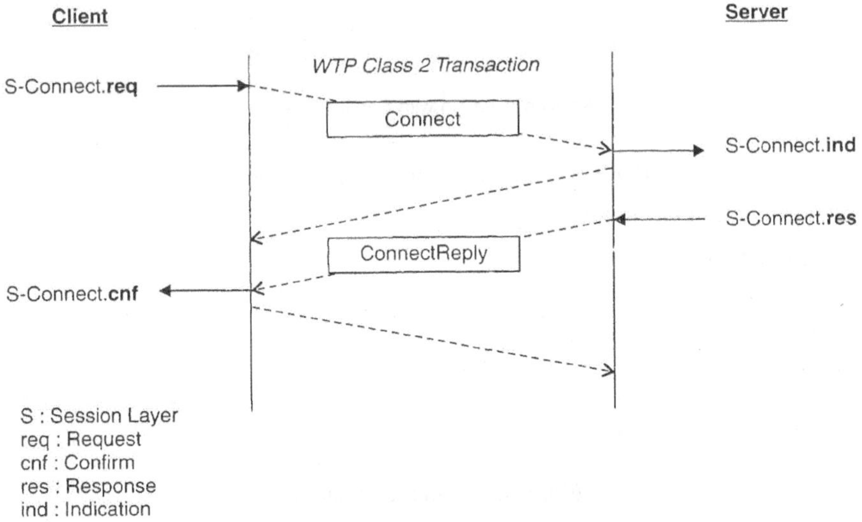

Fig. 5.6. Normal session creation

Session creation wherein the client is redirected to another server is shown in Fig. 5.7.

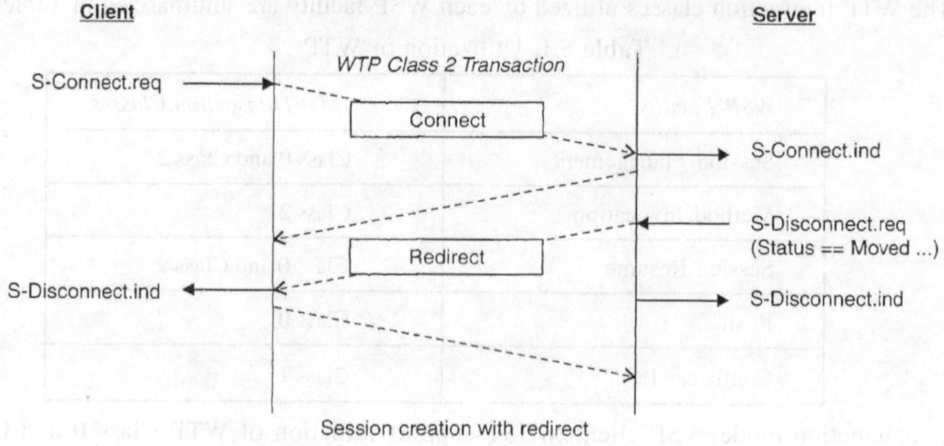

Fig. 5.7. Session creation with redirect

Session creation wherein the server session user refuses to accept the session is shown in Fig. 5.8.

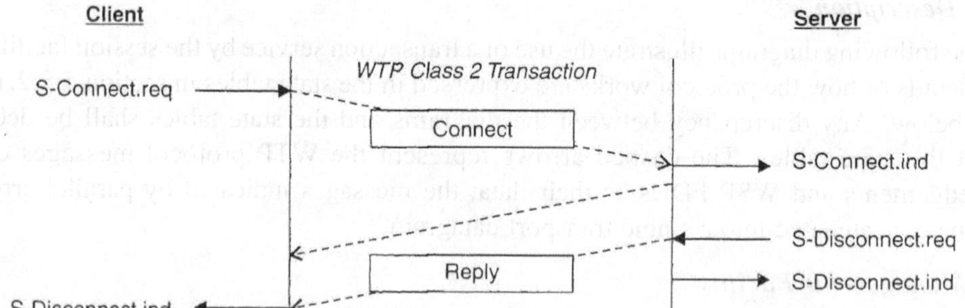

Fig. 5.8. Session creation with server error

Session termination is shown in Fig. 5.9.

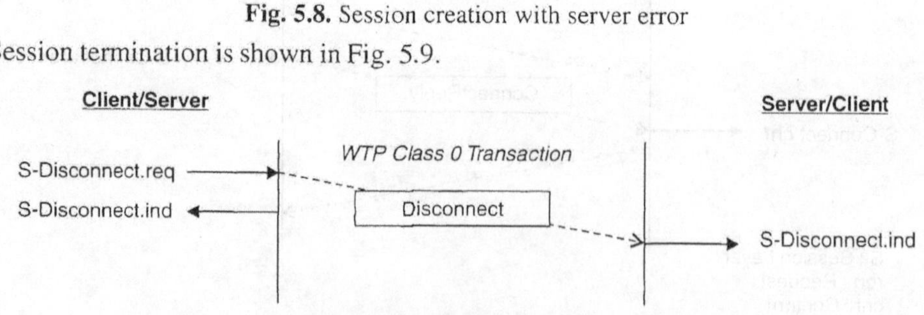

Fig. 5.9. Session termination

Session suspend is shown in Fig. 5.10.

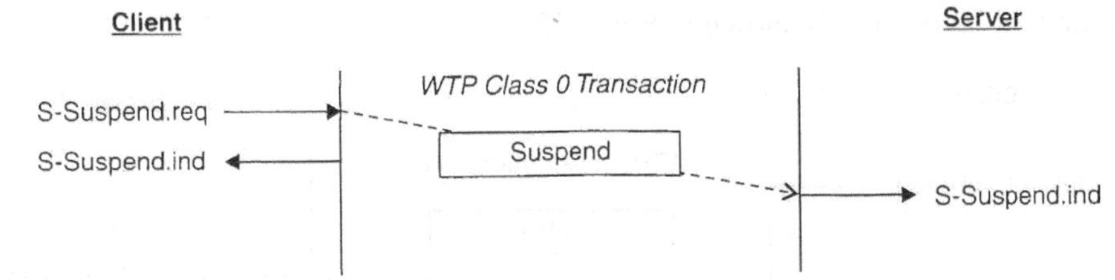

Fig. 5.10. Session suspend

When session resume succeeds, it proceeds as shown in Fig. 5.11.

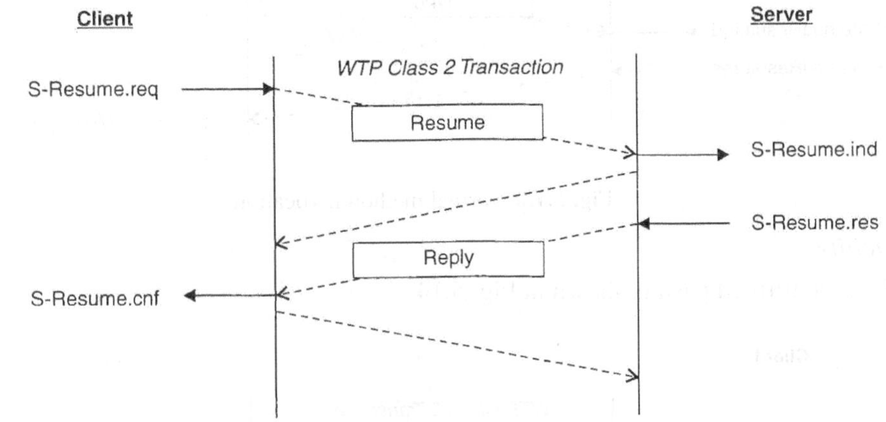

Fig. 5.11. Normal session resume

A session resume wherein the server session user refuses to resume the session is shown in Fig. 5.12.

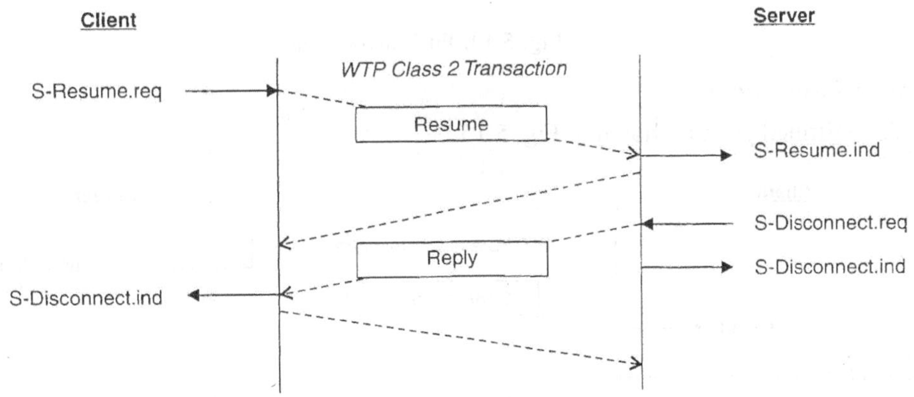

Fig. 5.12. Session resume with server error

Method Invocation Facility

A method invocation is shown in Fig. 5.13.

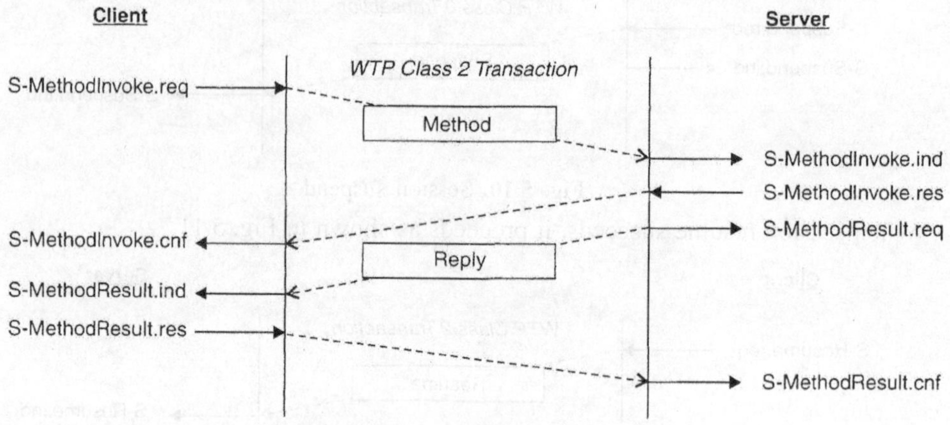

Fig. 5.13. Normal method invocation

Push Facility

An unconfirmed push is shown in Fig. 5.14.

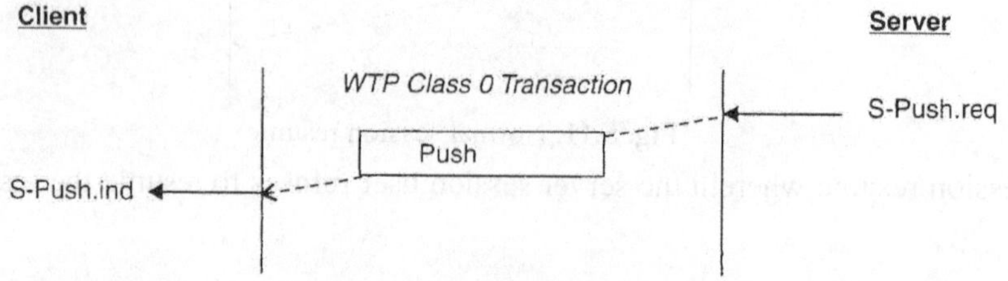

Fig. 5.14. Push invocation

Confirmed Push Facility

A confirmed push is shown in Fig. 5.15.

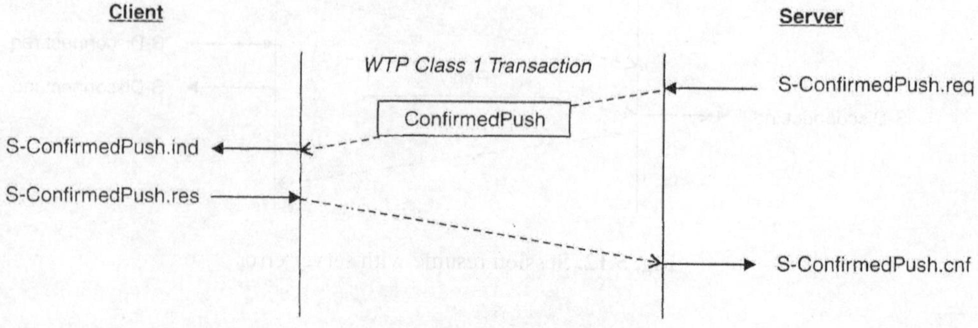

Fig. 5.15. Confirmed push invocation

Protocol Parameters

The protocol state machine uses the following parameters.

Maximum Receive Unit (MRU): The Maximum Receive Unit (MRU) is the size of the largest SDU the session layer can accept from the underlying service provider. The initial value is set to the default SDU sizes as specified in section 8.3.3, "Capability Defaults", below. The value can be modified during capability negotiation.

Maximum Outstanding Method Requests (MOM): The Maximum Outstanding Method Requests (MOM) is the number of method transactions that can be outstanding at a given time. The initial value is set to the default MOM as specified in section 8.3.3, "Capability Defaults", below. The value can be modified during capability negotiation.

Maximum Outstanding Push Requests (MOP): The Maximum Outstanding Push Requests (MOP) is the number of push transactions that can be outstanding at a given time. The value can be modified during capability negotiation.

Variables

The protocol state machine uses the following variables.

- *N_Methods*: N_Methods keeps track of the number of method transactions in process in the server.
- *N_Pushes*: N_Pushes keeps track of the number of push transactions in process in the client.
- *Session_ID*: Session_ID saves the session identifier assigned by the server in both the client and the server. The method used to assign the identifiers must be chosen so that a session identifier value cannot be repeated during the lifetime of a message in the used transport network; otherwise the session management logic may be confused.

Event Processing

Sessions are associated with a peer address quadruplet, *i.e.,* the client address, client port, server address, and server port. Incoming transactions are assigned to a particular session based on the peer address quadruplet. As a consequence, the peer address quadruplet is the true unique protocol-level identifier of a session. There can be only one session bound to a peer address quadruplet at a time.

In order to create a new session for a particular peer address quadruplet when one already appears to exist, the server session provider must allow for the creation of a *proto-session.* This is a second, constrained instance of a session that is used to process the session creation transaction on the server, *i.e.,* the Connect and Connect Reply PDUs; this is detailed in the table below.

State Tables

The state tables define the actions of connection-mode WSP. Because multiple methods and pushes can occur at the same time, there are three state tables defined for client and server: one for the session states, one for the states of a method and one for the states of a push. The state names used in the tables are logically completely separate from the states defined for the abstract service interface, although the names may be similar.

UNIT 6

WIRELESS LAN TECHNOLOGY

6.1 INTRODUCTION

The availability of wireless networking and wireless LANs can extend the freedom of a network user, solve various problems associated with hard-wired networks and even reduce network deployment costs in some cases. But, along with this freedom, wirelesses LANs bring a new set of challenges.

There are several wireless LAN solutions available today, with varying levels of standardization and interoperability. Two solutions that currently lead the industry are, HomeRF and Wi-Fi (IEEE 802.11*b*). Of these two, 802.11 technologies enjoy wider industry support and are targeted to solve Enterprise, Home and even public "hot spot" wireless LAN needs. The Wireless Ethernet Compatibility Alliance is working to provide certification of compliance with the 802.11 standards, helping to ensure multi-vendor interoperability.

Wide industry support for interoperability and operating system support address some of the deployment questions for wireless LANs. Still, wireless LANs present us with new challenges around security, roaming and configuration.

6.1.1 Wireless LAN Overview

High-speed wireless LANs can provide the benefits of network connectivity without the restrictions of being tied to a location or tethered by wires. There are many scenarios where this becomes interesting, including the following.

Wireless connections can extend or replace a wired infrastructure in situations where it is costly or prohibitive to lay cables. Temporary installations represent one example of when a wireless network might make sense or even be required. Some types of buildings or building codes may prohibit the use of wiring, making wireless networking an important alternative.

The increasingly mobile user becomes a clear candidate for a wireless LAN. Portable access to wireless networks can be achieved using laptop computers and wireless NICs. This enables the user to travel to various locations—meeting rooms, hallways, lobbies, cafeterias, classrooms, etc.

and still have access to their networked data. Without wireless access, the user would have to carry clumsy cabling and find a network tap to plug into.

Beyond the corporate campus, access to the Internet and even corporate sites could be made available through public wireless "hot spots" networks. Airports, restaurants, rail stations, and common areas throughout cities can be provisioned to provide this service. When the traveling worker reaches his or her destination, perhaps meeting a client at their corporate office, and limited access could be provided to the user through the local wireless network. The network can recognize the user from another corporation and create a connection that is isolated from local corporate network but provides Internet access to the visiting user.

In all these scenarios it is worth highlighting that today's standards-based wireless LANs operate at high speeds—the same speeds that were considered state of the art for wired networks just a few years ago. The access the user has is typically more than 11 megabits per second (Mbps) or about 30 to 100 times faster than standard dial up or Wireless WAN technologies. This bandwidth is certainly adequate to deliver a great user experience for a number of applications or services via the PC or mobile device. In addition, ongoing advancements with these wireless standards continue to increase bandwidth, with speeds of 22 Mbps.

6.2 TYPES OF WIRELESS LAN

Wireless LAN are generally categorized according to the transmission technique that is used, all current wireless LAN product fall into one of types. There are three types of LAN

- Infrared LANs
- Spread Spectrum LANs
- Narrowband microwave

6.2.1 Infrared LANs

An individual cell of an **IR LAN** is limited to a single room, because infrared light does not penetrate opaque walls. Will see some of characteristics details of infrared LANs.

- Strength and weaknesses
- Transmission techniques

Strength and Weaknesses

The infrared has two advantages; the infrared communication can be more easily secured against snooping than microwave and separate infrared installation. The strength of infrared is that the equipment is relatively inexpensive and simple. Infrared data transmission typically uses intensity modulation, so that IR receivers need to detect only the amplitude of optical signals, whereas most microwave receivers must detect frequency.

Infrared LANs have some drawbacks; many indoor environments experience rather intense infrared background radiation, from sunlight and indoor lighting. This ambient radiation appears as noise in an infrared receiver, requiring the use of transmitters of higher power than would otherwise be required and also limiting the range.

Transmission Techniques

There are three transmission techniques commonly used for IR data transmission.

There are 3 types of Infrared transmission techniques:

1. Directed beam infrared
2. Omni directional
3. Diffused

Directed Beam

- Directed-beam IR can be used to create point-to-point links:
  (*i*) Directed IR beams are used to create point to point links.
  (*ii*) A focused IR beam can have a range of several kilometers.
  (*iii*) Can be used for cross building interconnection or for indoor interconnection of LANs

Omni Directional

- Omni directional configuration involves a single base station that is with in line of sight of all other station on the LAN:
  (*i*) A single base station which is within the line of sight of all wireless stations on the LAN.
  (*ii*) Mounted on the ceiling
  (*iii*) It acts as a multiple repeater and broadcasts a omni directional beam.
  (*iv*) The wireless stations transmit a directional beam aimed at the ceiling base unit

Diffused

- Diffused configuration the IR transmitters are focused and aimed at a point on a diffusely reflecting ceiling:
  (*i*) All the IR transmitters are focused at a point on the reflecting ceiling.
  (*ii*) IR radiations' striking the ceiling is reradiated omni directionally and picked up by all the receivers in the area.

Advantages/Disadvantages

Infrared LANs Advantages:

1. Virtually unlimited bandwidth.
2. Unregulated spectrum.
3. IR light is diffusely reflected by light-colored objects. Can use this property to provide coverage in presence of obstacles.

4. Does not penetrate walls or other opaque objects.
5. Easy to secure.
6. Separate networks can operate without interference.

Infrared LANs Disadvantages

1. Interference from background radiation from sunlight.
2. Power restrictions due to eye safety.
3. Ambient radiation appears as noise in an infrared receiver.
4. Transmitters of higher power required.

6.2.2 Spread Spectrum LANs

This type of LAN makes use of spread spectrum transmission technology. In nearly all cases, these LANs operate in the ISM group so that no FCC licensing is required for their use in the US.

- Spread data transmission across the available frequency band in excess of minimum bandwidth required to send information.
- Makes it resistant to noise, interference, eavesdropping.
- Used with wireless LANs and other personal digital communication devices such as digital cellular phones

Configuration

In this LANs topology can be either hub or peer-to-peer. In the hub topology, the hub is usually swelled on the maximum and connected to a moral fiber wired LAN to provide connectivity. A peer-to-peer topology is one in which there is no hub. AMAC algorithm such as CSMA is used to control access. This topology is appropriate for **ad hoc LANs.**

Two of the most popular types of transmission using Spread Spectrum Analysis are:

1. Direct Sequence Spread Spectrum (DSSS)
2. Frequency Hopping Spread Spectrum (FSSS)

Direct Sequence Spread Spectrum (DSSS)

1. Most widely recognized form of Spread Spectrum.
2. Converts an incoming data bit stream into a symbol stream.
3. Each symbol represents a group of one or more bits.
4. Using QPKS techniques it multiplies each symbol with a noise like code called pseudorandom noise sequence.
5. Multiplication artificially increases used bandwidth based on the length of the chip sequence.

Frequency Hopping Spread Spectrum (FHSS)

1. Hops from narrow band to narrow band within a wide band.
2. The hopping pattern or sequence appears random but is actually a periodic sequence tracked by sender and receiver.
3. Can be susceptible to noise during any one hop but can achieve transmissions during other hops within the wideband.

Transmission Issues

The FCC has authorized two unlicensed application with in ISM band

- Which can operate at up to 1 watt in very low power system
- Which can operate at up to 0.5 watts, the FCC opened this band.

Advantages/Disadvantages

Spread Spectrum LANs Advantages:

1. Has the ability to eliminate or alleviate the effect of multipath interference.
2. Can share the same frequency band (overlay) with other users.
3. Provides privacy due to unknown random codes.
4. Involves low power spectral density since signal is spread over a large frequency band.

Spread Spectrum LANs Disadvantages:

1. The bandwidth is inefficient
2. The implementation of the Spread Spectrum LANs is somewhat complex

6.2.3 Narrowband microwave LANs

These LANs operate microwave frequencies but some of these products operate at frequencies that require FCC licensing. It has two-bandwidth they are:

- **Licensed Narrowband RF** has the advantages is that guarantees interference free communication.
- **Unlicensed Narrowband RF** can be used for narrowband transmission at lower power 0.5 watts or less.

Licensed Narrowband RF

1. Licensed within specific geographic areas to avoid potential interference.
2. Motorola - 600 licenses in 18-GHz range
    - Covers all metropolitan areas
    - Due to licensing, independent LANs in nearby locations don't interfere
    - Encrypted transmissions prevent eavesdropping guarantees interference free communication.

Unlicensed Narrowband RF

1. Radio LAN introduced narrowband wireless LAN in 1995.
2. Uses unlicensed ISM spectrum
3. Operates at 10 Mbps in the 5.8-GHz band
4. Range = 50 m to 100 m.

6.3 IEEE 802 PROTOCOL ARCHITECTURE

6.3.1 Introduction

IEEE 802 refers to a family of IEEE standards dealing with local area networks and networks. More specifically, the IEEE 802 standards are restricted to networks carrying variable-size packets. (By contrast, in cell-based networks data is transmitted in short, uniformly sized units called cells. Isochronous networks, where data is transmitted as a steady stream of octets, or groups of octets, at regular time intervals, are also out of the scope of this standard.) The number 802 was simply the next free number IEEE could assign, though "802" is sometimes associated with the date the first meeting was held-February 1980.

The services and protocols specified in IEEE 802 map to the lower two layers (Data Link and Physical) of the seven-layer OSI networking reference model. In fact, IEEE 802 splits the OSI Data Link Layer into two sub layers named Logical Link Control (LLC) and Media Access Control (MAC), so that the layers can be listed like this:

- Data link layer
  - (*i*) LLC sub layer
  - (*ii*) MAC sub layer
- Physical layer

The IEEE 802 family of standards is maintained by the IEEE 802 LAN/MAN Standards Committee (LMSC). The most widely used standards are for the Ethernet family, Token Ring, Wireless LAN, Bridging and Virtual Bridged LANs. An individual Working Group provides the focus for each area.

The table shows that various IEEE working groups:

Name	*Description*	*Note*
IEEE 802.1	Bridging (networking) and network management	
IEEE 802.2	Logical link control	Inactive
IEEE 802.3	Ethernet	
IEEE 802.4	Token bus	Disbanded
IEEE 802.5	Defines the MAC layer for a token ring	Inactive
IEEE 802.6	Metropolitan area networks	Disbanded
IEEE 802.7	Broadband LAN using coaxial cable	Disbanded
IEEE 802.8	Fiber optic TAG	Disbanded
IEEE 802.9	Integrated services LAN	Disbanded
IEEE 802.10	Interoperable LAN security	Disbanded
IEEE 802.11 a/b/g/n	Wireless LAN and mesh (Wi-Fi certification)	
IEEE 802.12	Demand priority	Disbanded
IEEE 802.13		Not used (officially)
IEEE 802.14	Cable modems	Disbanded
IEEE 802.15	Wireless PAN	
IEEE 802.15.1	Bluetooth certification	
IEEE 802.15.4	Zigbee certification	
IEEE 802.16	Broadband wireless access (WiMAX certification)	
IEEE 802.16e	(Mobile) broadband wireless access	
IEEE 802.16.1	Local multipoint distribution service	
IEEE 802.17	Resilient packet ring	
IEEE 802.18	Radio regulatory TAG	
IEEE 802.19	Coexistence TAG	
IEEE 802.20	Mobile broadband wireless access	
IEEE 802.21	Media independent handoff	
IEEE 802.22	Wireless regional area network	
IEEE 802.23	Broadband ISDN system	Experimental

6.3.2 IEEE 802 Architecture

IEEE 802 Protocol Layers Compared to OSI Model

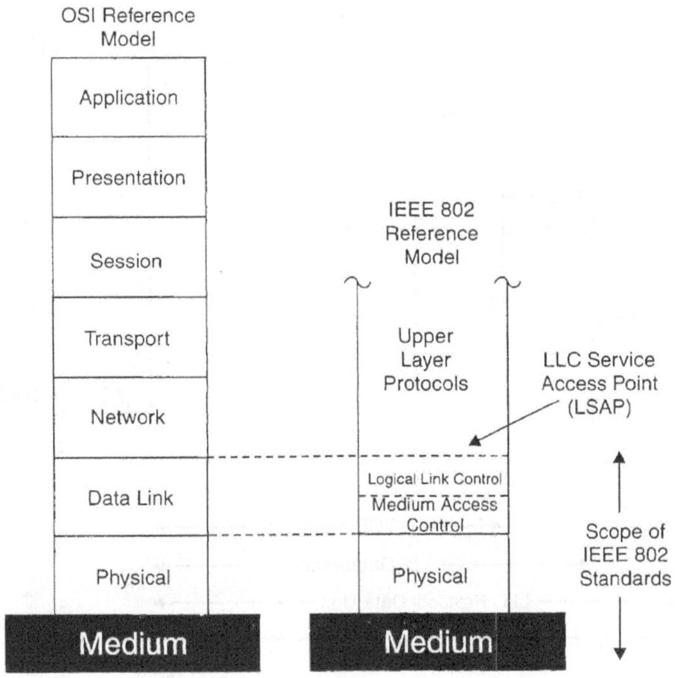

Fig. 6.1

6.3.3 Layers Functions

Physical Layer

- Encoding/decoding of signals
- Preamble generation/removal (for synchronization)
- Bit transmission/reception
- Specification of transmission medium and topology (considered below lowest layer of OSI model)

Medium Access Control (MAC) Layer

- On transmission, assemble data into a frame with address and error Detection fields
- On reception, disassemble frame, and perform address recognition and Error detection
- Govern access to the LAN transmission medium

Logical Link Control (LLC) Layer

- Provide an interface to higher layers and perform flow and error control
- Why the separation?

- Logic required to manage access to a shared-access medium is not found in traditional layer 2 data link control
- For the same LLC, several MAC options may be provided

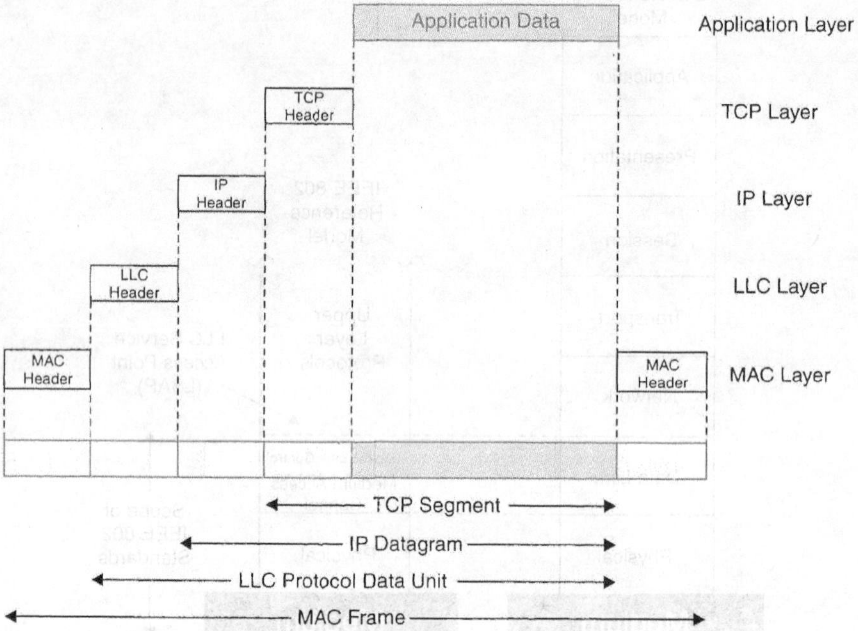

Fig. 6.2. IEEE 802 protocols in content

MAC Frame Format

(*i*) *MAC Control*: protocol control information needed for functioning of MAC protocol.

(*ii*) *Destination MAC Address*: destination physical attachment point on LAN.

(*iii*) *Source MAC Address*: source physical attachment point on LAN.

(*iv*) *Data*: body of MAC frame.

(*v*) *CRC*: cyclic redundancy check field (error detecting code).

- MAC layer is responsible for detecting errors and discarding any frames that are in error.
- LLC layer *optionally* keeps track of which frames have been successfully received and retransmits unsuccessful frames.
- Previous 2 tasks normally responsibility of data link protocol.
- LLC specifies mechanisms for addressing stations across the medium and for controlling the exchange of data between users.

LLC Services

Unacknowledged Connectionless Service: datagram-style service. No flow or error control mechanisms (delivery of data not guaranteed). How is reliability ensured then, if needed?

Connection-Mode Service: logic connection set up between 2 users, providing flow-control and error control.

Acknowledged Connectionless Service: datagram's to be acknowledged, but no prior logical connection is set up.

- MAC layer is responsible for detecting errors and discarding any Frames that are in error.

6.4 IEEE 802.11 ARCHITECTURE

6.4.1 Introduction

In 1990, IEEE 802.11 was formed with a charter to develop a MAC protocol and physical medium specifications. The difference between a portable and mobile station is that a portable station moves from point to point but is only used at a fixed point. Mobile stations access the LAN during movement.

When two or more stations come together to communicate with each other, they form a Basic Service Set (BSS). The minimum BSS consists of two stations. 802.11 LANs use the BSS as the standard building block.

A BSS that stands alone and is not connected to a base is called an Independent Basic Service Set (IBSS) or is referred to as an ad-hoc network. An ad-hoc network is a network where stations communicate only peer to peer. There is no base and no one gives permission to talk. Mostly these networks are spontaneous and can be set up rapidly. Ad-Hoc or IBSS networks are characteristically limited both temporally and spatially.

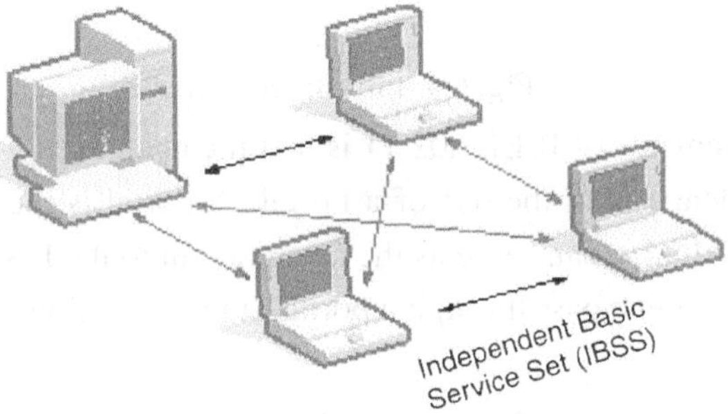

Fig. 6.3. Ad-hoc mode

When BSS's are interconnected the network becomes one with infrastructure. 802.11 infrastructures have several elements. Two or more BSS's are interconnected using a Distribution System or DS. This concept of DS increases network coverage.

Each BSS becomes a component of an extended, larger network. Entry to the DS is accomplished with the use of Access Points (AP). An access point is a station, thus addressable. So, data moves between the BSS and the DS with the help of these access points.

Creating large and complex networks using BSS's and DS's leads us to the next level of hierarchy, the Extended Service Set or ESS. The beauty of the ESS is the entire network looks like an independent basic service set to the Logical Link Control layer (LLC). This means that stations within the ESS can communicate or even move between BSS's transparently to the LLC.

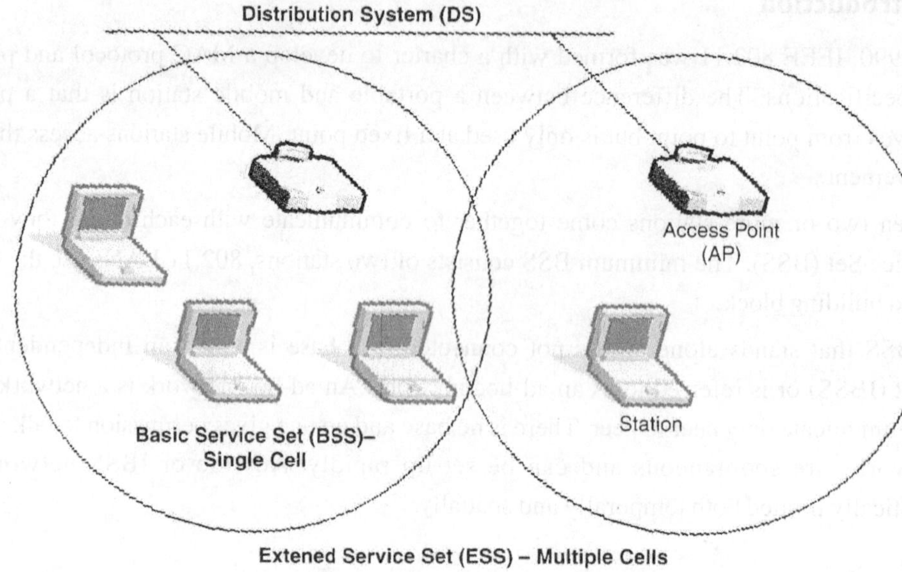

Fig. 6.4. Infrastructure mode

One of the requirements of IEEE 802.11 is that it can be used with existing wired networks. 802.11 solved this challenge with the use of a Portal. A portal is the logical integration between wired LANs and 802.11. It also can serve as the access point to the DS. All data going to an 802.11 LAN from an 802.X LAN must pass through a portal. It thus functions as bridge between wired and wireless.

The implementation of the DS is not specified by 802.11. Therefore, a distribution system may be created from existing or new technologies. A point-to-point bridge connecting LANs in two separate buildings could become a DS.

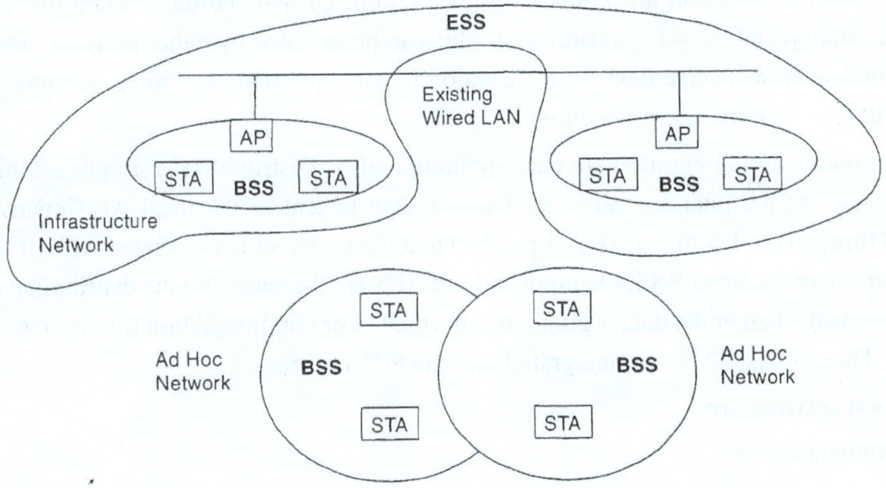

Fig. 6.5. Shows IEEE 802.11 architecture

6.4.2 Service of IEEE 802.11

While the implementation for the DS is not specified, 802.11 do specify the services, which the DS must support. Services are divided into two sections:

1. Station Services (SS)
2. Distribution System Services (DSS).

There are five services provided by the DSS:

1. Association
2. Reassociation
3. Disassociation
4. Distribution
5. Integration.

The first three services deal with station mobility. If a station is moving within its own BSS or is not moving, the stations mobility is termed No-transition. If a station moves between BSS's within the same ESS, its mobility is termed BSS-transition. If the station moves between BSS's of differing ESS's it is ESS transition. A station must affiliate itself with the BSS infrastructure if it wants to use the LAN. This is done by associating itself with an access point. Associations are dynamic in nature because stations move, turn on or turn off. A station can only be associated with one AP. This ensures that the DS always knows where the station is.

Association supports no-transition mobility but is not enough to support BSS-transition. Enter Reassociation. This service allows the station to switch its association from one AP to another. Both association and reassociation are initiated by the station. Disassociation is when the association between the station and the AP is terminated. This can be initiated by either party. A disassociated station cannot send or receive data. ESS-transition is not supported. A station can move to a new ESS but will have to reinitiate connections.

Distribution and Integration are the remaining DSS's. Distribution is simply getting the data from the sender to the intended receiver. The message is sent to the local AP (input AP), then distributed through the DS to the AP (output AP) that the recipient is associated with. If the sender and receiver are in the same BSS, the input and out AP's are the same. So the distribution service is logically invoked whether the data is going through the DS or not. Integration is when the output AP is a portal. Thus, 802.x LANs are integrated into the 802.11 DS.

Station services are:

1. Authentication
2. Deauthentication
3. Privacy
4. MAC Service Data Unit (MSDU) Delivery.

With a wireless system, the medium is not exactly bounded as with a wired system. In order to control access to the network, stations must first establish their identity. This is much like trying to enter a radio net in the military.

Two kinds of services:

(*i*) Basic Service Set (BSS)

(*ii*) Extended Service Set (ESS)

Basic Service Set (BSS)

- Made of stationary or mobile wireless stations and possible central Base stations (Access Point AP)
- Without an AP, a stand-alone network, cannot send data to the other BSSs (ad hoc architecture)
- Stations can form a network without the need of an AP (locate Each other be part of a BSS)

Extended Service Set (ESS)

- Two or more BSSs with APs
- BSSs connected through a distribution system (usually a wired LAN)
- Similar to a cellular network (a BSS is a cell and each AP a base station)
- MH can belong to more than one BSS at the same time
- ESS appears as a single LAN to LLC level.

6.4.3 IEEE 802.11 Physical Layer

The three physical layers originally defined in 802.11 included **two spread spectrum radio techniques** and **a diffuse infrared specification.**

The radio-based standards operate within the 2.4 GHz ISM band. These frequency bands are recognized by international regulatory agencies radio operations. As such, 802.11-based products do not require user licensing or special training.

Spread-spectrum techniques, in addition to satisfying regulatory requirements, increase reliability, boost throughput, and allow many unrelated products to share the spectrum without explicit cooperation and with minimal interference.

The original 802.11 wireless standard defines data rates of 1 Mbps and 2 Mbps via radio waves using Frequency Hopping Spread Spectrum (FHSS) or Direct Sequence Spread Spectrum (DSSS). It is important to note that FHSS and DSSS are fundamentally different signaling mechanisms and will not interoperate with one another.

Using the frequency hopping technique, the 2.4 GHz band is divided into 75 1-MHz sub channels. The sender and receiver agree on a hopping pattern, and data is sent over a sequence of the sub channels. Each conversation within the 802.11 network occurs over a different hopping pattern, and the patterns are designed to minimize the chance of two senders using the same sub channel simultaneously.

FHSS techniques allow for a relatively simple radio design, but are limited to speeds of no higher than 2 Mbps. This limitation is driven primarily by FCC (Federal Communications Commission USA) regulations that restrict sub channel bandwidth to 1 MHz. These regulations force FHSS systems to spread their usage across the entire 2.4 GHz band, meaning they must hop often, which leads to a high amount of hopping overhead.

In contrast, the direct sequence signaling technique divides the 2.4 GHz band into 14 22-MHz channels. Adjacent channels overlap one another partially, with three of the 14 being completely non-overlapping. Data is sent across one of these 22 MHz channels without hopping to other channels.

To compensate for noise on a given channel, a technique called "chipping" is used. Each bit of user data is converted into a series of redundant bit patterns called "chips." The inherent redundancy of each chip combined with spreading the signal across the 22 MHz channel provides for a form of error checking and correction; even if part of the signal is damaged, it can still be recovered in many cases, minimizing the need for retransmissions.

6.4.4 IEEE 802.11 Data Layer

The data link layer within 802.11 consists of two sub layers: **Logical Link Control (LLC)** and **Media Access Control (MAC).**

802.11 uses the same 802.2 LLC and 48-bit addressing as other 802 LANs, allowing for very simple bridging from wireless to IEEE wired networks, but the MAC is unique to WLANs.

The 802.11 MAC is very similar in concept to 802.3, in that it is designed to support multiple users on a shared medium by having the sender sense the medium before accessing it.

For 802.3 Ethernet LANs, the Carrier Sense Multiple Access with Collision Detection (CSMA/CD) protocol regulates how Ethernet stations establish access to the wire and how they detect and handle collisions that occur when two or more devices try to simultaneously communicate over the LAN. In an 802.11 WLAN, collision detection is not possible due to what is known as the "near/far" problem: to detect a collision, a station must be able to transmit and listen at the same time, but in radio systems the transmission drowns out the ability of the station to "hear" a collision.

Medium Access Control

Functionality

Reliable data delivery

Fairly control access

Protection of data

Deals

Noisy and unreliable medium

Frame exchange protocol - ACK

Overhead to IEEE 802.3 -

Hidden Node Problem - RTS/CTS

Participation of all stations

Reaction to every frame

Basic Access Mechanism

CSMA/CA

Binary exponential back-off

NAV - Network Allocation Vector

To account for this difference, 802.11 use a slightly modified protocol known as Carrier Sense Multiple Access with Collision Avoidance (CSMA/CA) or the Distributed Coordination Function (DCF). CSMA/CA attempts to avoid collisions by using explicit packet Acknowledgment (ACK), which means an ACK packet is sent by the receiving station to confirm that the data packet arrived intact.

CSMA/CA works as follows. A station wishing to transmit senses the air, and, if no activity is detected, the station waits an additional, randomly selected period of time and then transmits if the medium is still free. If the packet is received intact, the receiving station issues an ACK frame that, once successfully received by the sender, completes the process. If the ACK frame is not detected by the sending station, either because the original data packet was not received intact or the ACK was not received intact, a collision is assumed to have occurred and the data packet is transmitted again after waiting another random amount of time.

CSMA/CA thus provides a way of sharing access over the air. This explicit ACK mechanism also handles interference and other radio related problems very effectively. However, it does add some overhead to 802.11 that 802.3 does not have, so that an 802.11 LAN will always have slower performance than an equivalent Ethernet LAN.

Another MAC-layer problem specific to wireless is the "hidden node" issue, in which two stations on opposite sides of an access point can both "hear" activity from an access point, but not from each other, usually due to distance or an obstruction.

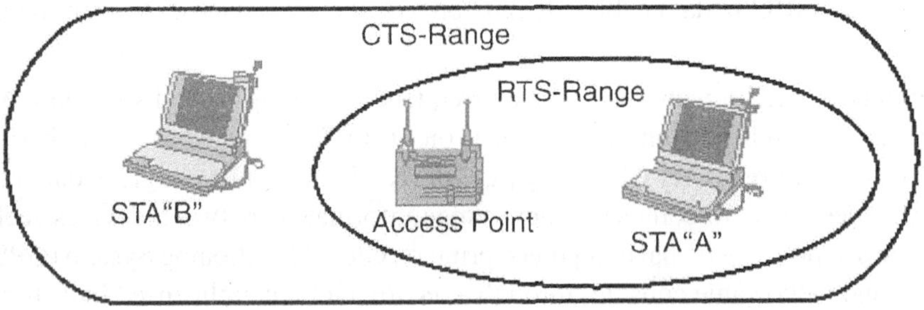

Fig. 6.6. RTS/CTS procedure eliminates the "hidden node" problem

To solve this problem, 802.11 specify an optional Request to Send/Clear to Send (RTS/CTS) protocol at the MAC layer. When this feature is in use, a sending station transmits an RTS and waits for the access point to reply with a CTS. Since all stations in the network can hear the access point, the CTS causes them to delay any intended transmissions, allowing the sending station to transmit and receive a packet acknowledgment without any chance of collision.

Since RTS/CTS adds additional overhead to the network by temporarily reserving the medium, it is typically used only on the largest-sized packets, for which retransmission would be expensive from a bandwidth standpoint.

Finally, the 802.11 MAC layer provides for two other robustness features: CRC checksum and packet fragmentation. Each packet has a CRC checksum calculated and attached to ensure that the data was not corrupted in transit. This is different from Ethernet, where higher-level protocols such as TCP handle error checking. Packet fragmentation allows large packets to be broken into smaller units when sent over the air, which is useful in very congested environments or when interference is a factor, since larger packets have a better chance of being corrupted. This technique reduces the need for retransmission in many cases and thus improves overall wireless network performance. The MAC layer is responsible for reassembling fragments received, rendering the process transparent to higher level protocols.

UNIT 7 BLUETOOTH

7.1 INTRODUCTION TO BLUETOOTH

Bluetooth is an open wireless protocol for exchanging data over short distances from fixed and mobile devices, creating Personal Area Networks (PANs). It was originally conceived as a wireless alternative to RS232 data cables. It can connect several devices, overcoming problems of synchronization.

Bluetooth uses a radio technology called frequency-hopping spread spectrum, which chops up the data being sent and transmits chunks of it on up to 79 frequencies. In its basic mode, the modulation is Gaussian Frequency-Shift Keying (GFSK). It can achieve a gross data rate of 1 Mb/s. Bluetooth provides a way to connect and exchange information between devices such as mobile phones, telephones, laptops, personal computers, printers, Global Positioning System (GPS) receivers, digital cameras, and video game consoles through a secure, globally unlicensed Industrial, Scientific and Medical (ISM) 2.4 GHz short-range radio frequency bandwidth. The Bluetooth specifications are developed and licensed by the Bluetooth Special Interest Group (SIG). The Bluetooth SIG consists of companies in the areas of telecommunication, computing, networking, and consumer electronics.

Bluetooth is a standard and communications protocol primarily designed for low power consumption, with a short range (power-class-dependent: 1 meter, 10 meters, 100 meters) based on low-cost transceiver microchips in each device. Bluetooth makes it possible for these devices to communicate with each other when they are in range. Because the devices use a radio (broadcast) communications system, they do not have to be in line of sight of each other.

Class	*Maximum Permitted Power mW (dBm)*	*Range (approximate)*
Class 1	100 mW (20 dBm)	~100 meters
Class 2	2.5 mW (4 dBm)	~10 meters
Class 3	1 mW (0 dBm)	~1 meter

In most cases the effective range of class 2 devices is extended if they connect to a class 1 transceiver, compared to a pure class 2 network. This is accomplished by the higher sensitivity and transmission power of Class 1 devices.

Version	*Data Rate*
Version 1.2	1 Mbit/s
Version 2.0 + EDR	3 Mbit/s
WiMedia Alliance (proposed)	53-480 Mbit/s

7.1.1 Technology Comparison

Since Bluetooth operates in the unlicensed ISM band that is also used by other devices such as 802.11 networks, baby monitors, garage door openers, microwave ovens etc, there is possibility of interference. Bluetooth uses Frequency Hop Spread Spectrum (FHSS) to avoid any interference. A Bluetooth channel is divided into time slots each 625 micro second in length. The devices hop through these timeslots making 1600 hops per second. This trades bandwidth efficiency for reliability, integrity and security.

Category	*Home-RF (1.09)*	*802.11*	*Bluetooth*	*IrDA (AIR)*
Market	Home WLAN	WLAN	Cable	Cable
Technology	RF : 2.4 GHz FHSS	RF : 24 GHz FHSS/DSSS	RF : 2.4 GHz FHSS	Optical 850 nm
Power	20dBm	20dBm	0/20dBm	?
Symbol Rate	0.8/1.6 M	11M	1 M	4M/115K
Distance	50 m	30-100 m	0-10 m/100 m	0-3m/5m
Topology	128 devices CSMA	128 devices CSMA	8 devices Pt to MP	10 devices Pt to MP
Security	Optional	Optional WEP	Authentication, Key, mgmt, Encryption	Application Layer
Cost	Low	High	Low	Low

Fig. 7.1. Technology comparison

7.2 BLUETOOTH ARCHITECTURE

Bluetooth communication occurs between a master radio and a slave radio. Bluetooth radios are symmetric in that the same device may operate as a master and also the slave. Each radio has a 48 bit unique device address (BD_ADDR) that is fixed. Two or more radio devices together form ad-hoc networks called piconets. All units within a piconet share the same channel. Each piconet has one master device and one or more slaves. There may be up to seven active slaves at a time within a piconet. Thus each active device within a piconet is identifiable by a 3 bit active device address. Inactive slaves in unconnected modes may continue to reside within the piconet. A master is the only one that may initiate a Bluetooth communication link. However, once a link is established, the slave may request a master/slave switch to become the master. Slaves are not allowed to talk to each other directly. All communication occurs within the slave and the master. Slaves within a piconet must also synchronize their internal clocks and frequency hops with that of the master. Each piconet uses a different frequency hopping sequence. Radio devices used Time Division Multiplexing (TDM). A master device in a piconet transmits on even numbered slots and the slaves may transmit on odd numbered slots. Multiple piconets with overlapping coverage areas form a scatternet. Each piconet may have only one master, but slaves may participate in different piconets on a time-division multiplex basis. A device may be a master in one piconet and a slave in another or a slave in more than one piconet.

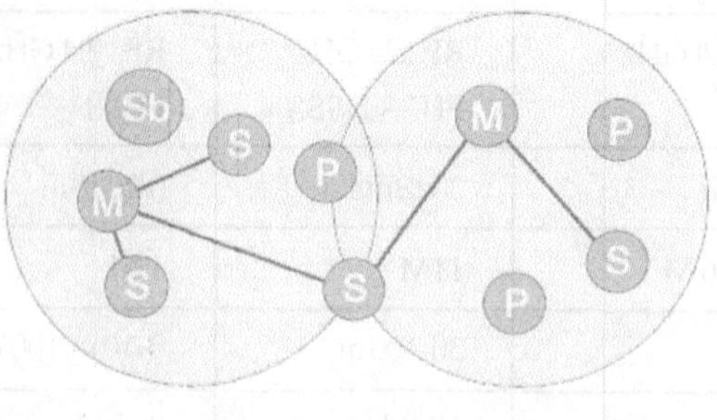

Fig. 7.2. A Scatternet comprising two piconets

7.3 THE BLUETOOTH PROTOCOL STACK

"Bluetooth is defined as a layer protocol architecture consisting of core protocols, cable replacement protocols, telephony control protocols, and adopted protocols".

The protocols of the Bluetooth stack can be divided into four different categories:

Bluetooth Core Protocols	Baseband, LMP, L2CAP, SDP
Cable Replacement Protocol	RFCOMM
Telephony Control Protocol	TCS Binary, AT Commands
Adopted Protocols	PPP, TCP/IP, OBEX, WAP, vCard, vCal, IrMC, WAE

Mandatory protocols for all Bluetooth stacks are: **BASEBAND, LMP, L2CAP and SDP**

Additionally, these protocols are almost universally supported: **HCI and RFCOMM.**

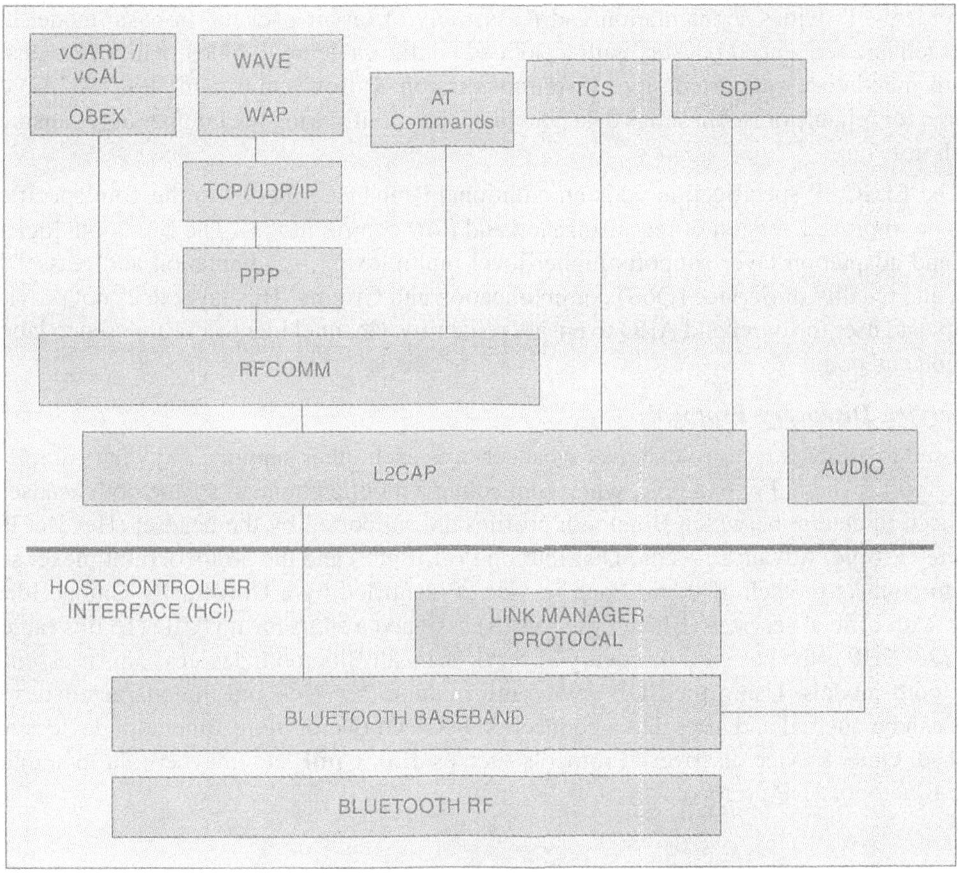

Fig. 7.3. The bluetooth protocol stack

7.3.1 Bluetooth Core Protocols

Baseband

The baseband and the Link control layers enable the physical RF link between Bluetooth devices to form a piconet. Both circuit and packet switching is used. They provide two kinds of physical links using the baseband packets. Synchronous Connection Oriented (SCO) and Asynchronous connectionless (ACL). ACL packets are used for data only, while the SCO packets may contain audio only or a combination of audio and data.

LMP (Link Management Protocol)

Used for control of the radio link between two devices. Implemented on the controller. The link manager protocol is responsible for the link set up between Bluetooth units. This protocol layer caters to issues of security like authentication, encryption by generating, exchanging and checking the link and encryption keys. It also deals with control and negotiation of baseband packet sizes.

L2CAP (Logical Link Control and Adaptation Protocol)

Used to multiplex multiple logical connections between two devices using different higher level protocols. Provides segmentation and reassembly of on-air packets. In basic mode, L2CAP provides reliable sequenced packets with a payload configurable up to 64kB, with 672 bytes as the minimum mandatory supported size. In retransmission & flow control modes, L2CAP can be configured for reliable or isochronous data per channel by configuring the number of retransmissions and flush time out.

The EL2CAP specification adds an additional "enhanced mode" to the core specification, which is an improved version of retransmission and flow control modes. The Bluetooth logical link control and adaptation layer supports higher level multiplexing, segmentation and reassembly of packets, and Quality of Service (QoS) communication and Groups. This layer does not provide any reliability and uses the baseband ARQ to ensure reliability. Channel identifiers are used to label each connection end point.

SDP (Service Discovery Protocol)

Used to allow devices to discover what services each other support, and what parameters to use to connect to them. For example, when connecting a mobile phone to a Bluetooth headset, SDP will be used to determine which Bluetooth profiles are supported by the headset (Headset Profile, Hands Free Profile, Advanced Audio Distribution Profile etc.) and the protocol multiplexer settings needed to connect to each of them. Each service is identified by a Universally Unique Identifier (UUID), with official services (Bluetooth profiles) assigned a short form UUID (16 bits rather than the full 128) SDP is the basis for discovery of services on all Bluetooth devices. This is essential for all Bluetooth models. Using the SDP device information, services and the characteristics of the services can be queried and after that a connection between two or more Bluetooth devices may be established. Other service discovery Protocols such as **Jini, UpnP** etc. maybe used in conjunction with the Bluetooth SDP protocol.

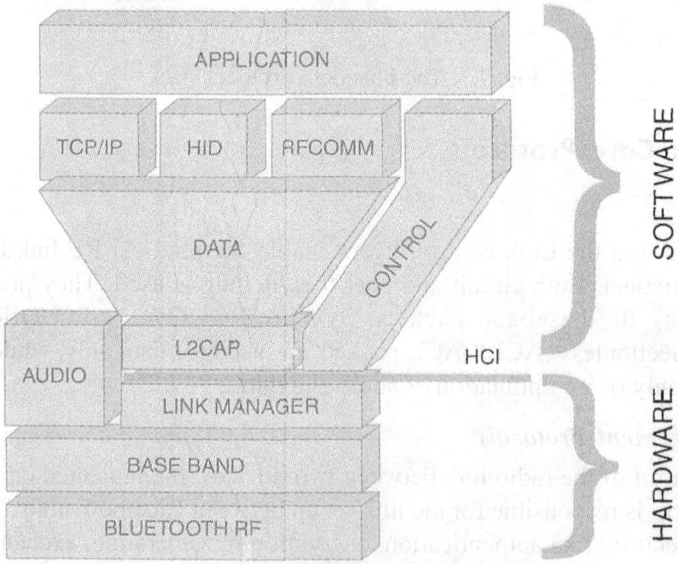

Fig. 7.4. Core protocols

7.3.2 HCI (Host/Controller Interface)

Standardized communication between the host stack (*e.g.* a PC or mobile phone OS) and the controller (the Bluetooth I.C.) This standard allows the host stack or controller I.C. to be swapped with minimal adaptation.

There are several HCI transport layer standards, each using a different hardware interface to transfer the same command, event and data packets. The most commonly used are USB (in PCs) and UART (in mobile phones and PDAs).

In Bluetooth devices with simple functionality, *e.g.* headsets, the host stack and controller can be implemented on the same microprocessor. In this case the HCI is optional, although often implemented as an internal software interface.

7.3.3 RFCOMM (Cable Replacement Protocol)

Radio frequency communications (RFCOMM) is the cable replacement protocol used to create a virtual serial data stream. RFCOMM provides for binary data transport and emulates EIA-232 (formerly RS-232) control signals over the Bluetooth baseband layer.

RFCOMM provides a simple reliable data stream to the user, similar to TCP. It is used directly by many telephony related profiles as a carrier for AT commands, as well as being a transport layer for OBEX over Bluetooth.

Many Bluetooth applications use RFCOMM because of its widespread support and publicly available API on most operating systems. Additionally, applications that used a serial port to communicate can be quickly ported to use RFCOMM.

BNEP (Bluetooth Network Encapsulation Protocol)

BNEP is used to transfer another protocol stack's data via an L2CAP channel. Its main purpose is the transmission of IP packets in the Personal Area Networking Profile. BNEP performs a similar function to SNAP in Wireless LAN.

AVCTP (Audio/Visual Control Transport Protocol)

Used by the remote control profile to transfer AV/C commands over an L2CAP channel. The music control buttons on a stereo headset use this protocol to control the music player.

AVDTP (Audio/Visual Data Transport Protocol)

Used by the advanced audio distribution profile to stream music to stereo headsets over an L2CAP channel. Intended to be used by video distribution profile.

7.3.4 Telephone Control Protocol

Telephony control protocol-binary (**TCS BIN**) is the bit-oriented protocol that defines the call control signaling for the establishment of voice and data calls between Bluetooth devices. Additionally, "TCS BIN defines mobility management procedures for handling groups of Bluetooth TCS devices".

TCS-BIN is only used by the cordless telephony profile, which failed to attract implementers. As such it is only of historical interest.

7.3.5 Adopted Protocols

Adopted protocols are defined by other standards-making organizations and incorporated into Bluetooth's protocol stack, allowing Bluetooth to create protocols only when necessary. The adopted protocols include:

- *Point-to-Point Protocol (PPP)*: Internet standard protocol for transporting IP datagram's over a point-to-point link.
- *TCP/IP/UDP*: Foundation Protocols for TCP/IP protocol suite.
- *Object Exchange Protocol (OBEX)*: Session-layer protocol for the exchange of objects, providing a model for object and operation representation.
- *Wireless Application Environment / Wireless Application Protocol (WAE/WAP)*: WAE specifies an application framework for wireless devices and WAP is an open standard to provide mobile users access to telephony and information services.

7.4 BLUETOOTH FORMATS

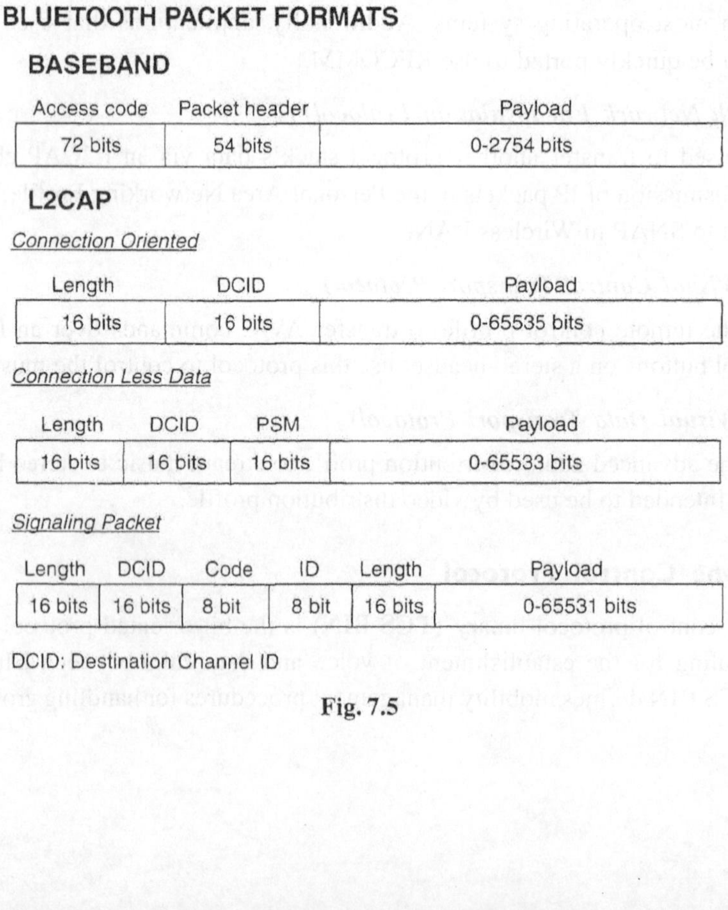

Fig. 7.5

7.5 APPLICATIONS OF BLUETOOTH

More prevalent applications of Bluetooth include:

- Wireless control of and communication between a mobile phone and a hands-free headset. This was one of the earliest applications to become popular.
- Wireless networking between PCs in a confined space and where little bandwidth is required.
- Wireless communication with PC input and output devices, the most common being the mouse, keyboard and printer.
- Transfer of files, contact details, calendar appointments, and reminders between devices with OBEX.
- Replacement of traditional wired serial communications in test equipment, GPS receivers, medical equipment, bar code scanners, and traffic control devices.
- For controls where infrared was traditionally used.
- For low bandwidth applications where higher [USB] bandwidth is not required and cable-free connection desired.
- Sending small advertisements from Bluetooth-enabled advertising hoardings to other, discoverable, Bluetooth devices.
- Two seventh-generation game consoles, Nintendo's Wii and Sony's PlayStation 3, use Bluetooth for their respective wireless controllers.
- Dial-up internet access on personal computers or PDAs using a data-capable mobile phone as a modem.

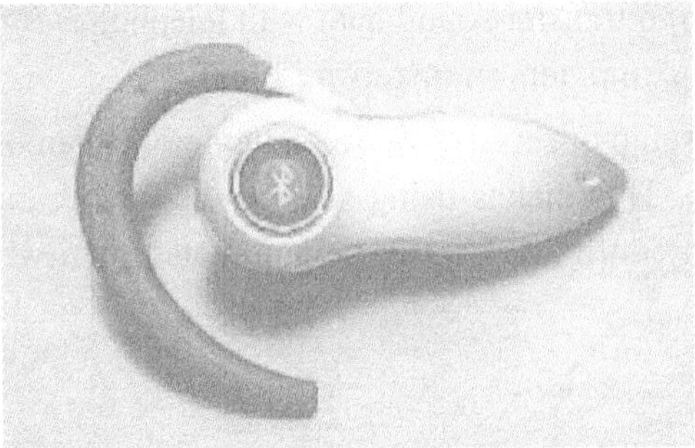

A typical Bluetooth mobile phone headset

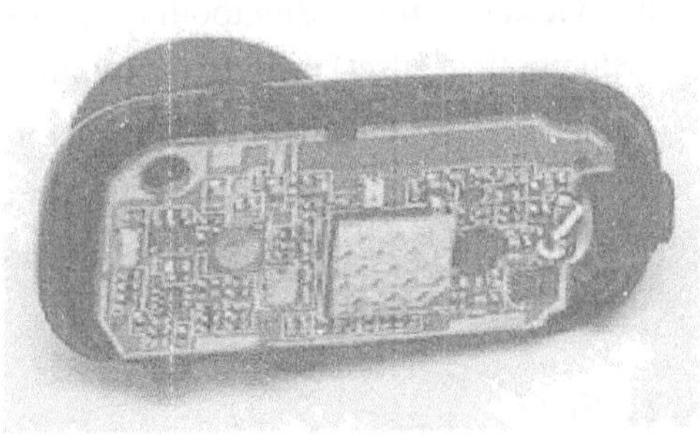

Nokia BH-208 headset internals

Fig. 7.6

7.6 BLUETOOTH IEEE 802.15.1 VS. WI-FI IEEE 802.11 IN NETWORKING

Bluetooth and Wi-Fi have many applications in today's offices, homes, and on the move: setting up networks, printing, or transferring presentations and files from PDAs to computers. Both are versions of unlicensed wireless technology.

Wi-Fi is intended for resident equipment and its applications. The category of applications is outlined as WLAN, the wireless local area networks. Wi-Fi is intended as a replacement for cabling for general local area network access in work areas.

Bluetooth is intended for non resident equipment and its applications. The category of applications is outlined as the Wireless Personal Area Network (WPAN). Bluetooth is a replacement for cabling in a variety of personally carried applications in any ambience.

7.6.1 Bluetooth Devices

Fig. 7.7. A Bluetooth USB dongle with a 100 m range

Bluetooth exists in many products, such as telephones, the Wii, PlayStation 3, and Lego Mindstorms NXT and recently in some high definition watches, modems and headsets. The technology is useful when transferring information between two or more devices that are near each other in low-bandwidth situations. Bluetooth is commonly used to transfer sound data with telephones (*i.e.*, with a Bluetooth headset) or byte data with hand-held computers (transferring files).

Bluetooth protocols simplify the discovery and setup of services between devices. Bluetooth devices can advertise all of the services they provide. This makes using services easier because more of the security, network address and permission configuration can be automated than with many other network types.

7.6.2 Wi-Fi

Wi-Fi is more like a traditional Ethernet network, and requires configuration to set-up shared resources, transmit files, and to set up audio links (for example, headsets and hands-free devices). Wi-Fi uses the same radio frequencies as Bluetooth, but with higher power, resulting in a stronger connection. Wi-Fi is sometimes called "wireless ethernet." This description is accurate, as it also provides an indication of its relative strengths and weaknesses. Wi-Fi requires more set-up but is better suited for operating full-scale networks; it enables a faster connection, better range from the base station, and better security than Bluetooth.

7.7 COMPUTER REQUIREMENTS

A personal computer must have a Bluetooth adapter in order to communicate with other Bluetooth devices (such as mobile phones, mice and keyboards). While some desktop computers and most recent laptops come with a built-in Bluetooth adapter, others will require an external one in the form of a dongle.

Unlike its predecessor, IrDA, which requires a separate adapter for each device, Bluetooth allows multiple devices to communicate with a computer over a single adapter.

A typical Bluetooth USB dongle

An internal notebook Bluetooth card (14×36×4 mm)

Fig. 7.8

7.7.1 Operating System Support

Apple has supported Bluetooth since Mac OS X v10.2 which was released in 2002.

For Microsoft platforms, Windows XP Service Pack 2 and later releases have native support for Bluetooth. Previous versions required users to install their Bluetooth adapter's own drivers, which were not directly supported by Microsoft. Microsoft's own Bluetooth dongles (packaged with their Bluetooth computer devices) have no external drivers and thus require at least Windows XP Service Pack 2.

Linux has two popular Bluetooth stacks, BlueZ and Affix. The BlueZ stack is included with most Linux kernels and was originally developed by Qualcomm. The Affix stack was developed by Nokia. FreeBSD features Bluetooth support since its 5.0 release. NetBSD features Bluetooth support since its 4.0 release. Its Bluetooth stack has been ported to OpenBSD as well.

7.8 MOBILE PHONE REQUIREMENTS

A mobile phone that is Bluetooth enabled is able to pair with many devices. To ensure the broadest support of feature functionality together with legacy device support, the Open Mobile Terminal Platform (OMTP) forum given some rules for mobile phone requirements.

7.9 SPECIFICATIONS AND FEATURES

The Bluetooth specification was developed in 1994 by Jaap Haartsen and Sven Mattisson, who were working for Ericsson Mobile Platforms in Lund, Sweden.The specification is based on frequency-hopping spread spectrum technology.

The specifications were formalized by the Bluetooth Special Interest Group (SIG). The SIG was formally announced on May 20, 1998. Today it has a membership of over 11,000 companies worldwide. It was established by Ericsson, IBM, Intel, Toshiba, and Nokia, and later joined by many other companies.

Bluetooth 1.0 and 1.0B

Versions 1.0 and 1.0B had many problems, and manufacturers had difficulty making their products interoperable. Versions 1.0 and 1.0B also included mandatory Bluetooth hardware device address (BD_ADDR) transmission in the Connecting process (rendering anonymity impossible at the protocol level), which was a major setback for certain services planned for use in Bluetooth environments.

Bluetooth 1.1

- Ratified as IEEE Standard 802.15.1-2002.
- Many errors found in the 1.0B specifications were fixed.
- Added support for non-encrypted channels.
- Received Signal Strength Indicator (RSSI).

Bluetooth 1.2

This version is backward compatible with 1.1 and the major enhancements include the following:

- Faster Connection and Discovery
- Adaptive Frequency-Hopping spread spectrum (AFH), which improves resistance to radio frequency interference by avoiding the use of crowded frequencies in the hopping sequence.
- Higher transmission speeds in practice, up to 721 kbit/s, than in 1.1.
- Extended Synchronous Connections (ESCO), which improve voice quality of audio links by allowing retransmissions of corrupted packets, and may optionally increase audio latency to provide better support for concurrent data transfer.
- Host Controller Interface (HCI) support for three-wire UART.
- Ratified as IEEE Standard 802.15.1-2005.
- Introduced Flow Control and Retransmission Modes for L2CAP.

Bluetooth 2.0

This version of the Bluetooth specification was released on November 10, 2004. It is backward compatible with the previous version 1.2. The main difference is the introduction of an Enhanced Data Rate (EDR) for faster data transfer. The nominal rate of EDR is about 3 megabits per second, although the practical data transfer rate is 2.1 megabits per second. The additional throughput is obtained by using a different radio technology for transmission of the data. Standard, or Basic Rate, transmission uses Gaussian Frequency Shift Keying (GFSK) modulation of the radio signal with a gross air data rate of 1 Mbit/s. EDR uses a combination of GFSK and Phase Shift Keying modulation (PSK) with two variants, $\pi/4$-DQPSK and 8DPSK. These have gross air data rates of 2, and 3 Mbit/s respectively.

According to the 2.0 specification, EDR provides the following benefits:

- Three times faster transmission speed—up to 10 times (2.1 Mbit/s) in some cases.
- Reduced complexity of multiple simultaneous connections due to additional bandwidth.
- Lower power consumption through a reduced duty cycle.

The Bluetooth Special Interest Group (SIG) published the specification as "Bluetooth 2.0 + EDR" which implies that EDR is an optional feature. Aside from EDR, there are other minor improvements to the 2.0 specification, and products may claim compliance to "Bluetooth 2.0" without supporting the higher data rate. At least one commercial device, the HTC TyTN Pocket PC phone, states "Bluetooth 2.0 without EDR" on its data sheet.

Bluetooth 2.1

Bluetooth Core Specification Version 2.1 is fully backward compatible with 1.2, and was adopted by the Bluetooth SIG on July 26, 2007.This specification includes the following features:

Extended Inquiry Response: Provides more information during the inquiry procedure to allow better filtering of devices before connection. This information may include the name of the device, a list of services the device supports and the transmission power level used for inquiry responses.

Sniff Sub Rating: Reduces the power consumption when devices are in the sniff low-power mode, especially on links with asymmetric data flows. Human Interface Devices (HID) are expected to benefit the most, with mouse and keyboard devices increasing their battery life by a factor of 3 to 10. It lets devices decide how long they will wait before sending keep alive messages to one another. Previous Bluetooth implementations featured keep alive message frequencies of up to several times per second. In contrast, the 2.1 specification allows pairs of devices to negotiate this value between them to as infrequently as once every 5 or 10 seconds.

Encryption Pause Resume: Enables an encryption key to be refreshed, enabling much stronger encryption for connections that stay up for longer than 23.3 hours (one Bluetooth day).

Secure Simple Pairing: Radically improves the pairing experience for Bluetooth devices, while increasing the use and strength of security. It is expected that this feature will significantly increase the use of Bluetooth.

Near Field Communication (NFC) Cooperation: Automatic creation of secure Bluetooth connections when NFC radio interface is also available. This functionality is part of the Secure Simple Pairing where NFC is one way of exchanging pairing information. For example, a headset should be paired with a Bluetooth 2.1 phone including NFC just by bringing the two devices close to each other (a few centimeters). Another example is automatic uploading of photos from a mobile phone or camera to a digital picture frame just by bringing the phone or camera close to the frame.

Bluetooth 3.0

The 3.0 specification was adopted by the Bluetooth SIG on April 21st, 2009. Its main new feature is AMP (Alternate MAC/PHY), the addition of 802.11 as a high speed transport. Two technologies had been anticipated for AMP: 802.11 and UWB, but UWB is missing from the specification.

- *Alternate MAC PHY*: Enables the use of alternative MAC and PHY's for transporting Bluetooth profile data. The Bluetooth Radio is still used for device discovery, initial connection and profile configuration, however when lots of data needs to be sent, the high speed alternate MAC PHY (802.11, typically associated with Wi-Fi) will be used to transport the data. This means that the proven low power connection models of Bluetooth are used when the system is idle, and the low power per bit radios are used when lots of data needs to be sent.
- *Unicast Connectionless Data*: Permits service data to be sent without establishing an explicit L2CAP channel. It is intended for use by applications that require low latency between user action and reconnection/transmission of data. This is only appropriate for small amounts of data.

7.10 EXTRA INFORMATION OF BLUETOOTH

7.10.1 Bluetooth Low Energy

On April 20, 2009, Bluetooth SIG presented the new Bluetooth low energy as an entirely additional protocol stack, compatible with other existing Bluetooth protocol stacks. The preceding naming as 'Wibree' and 'Bluetooth ULP' (Ultra Low Power) has been outdated by the final naming as 'Bluetooth low energy'.

On June 12, 2007, Nokia and Bluetooth SIG had announced that Wibree will be a part of the Bluetooth specification, as an ultra-low power Bluetooth technology. Expected use cases include watches displaying Caller ID information, sports sensors monitoring your heart rate during exercise, and medical devices. The Medical Devices Working Group is also creating a medical devices profile and associated protocols to enable this market. Bluetooth low energy technology is designed for devices to have a battery life of up to one year.

Future

- *Broadcast Channel*: Enables Bluetooth information points. This will drive the adoption of Bluetooth into mobile phones, and enable advertising models based around users pulling information from the information points, and not based around the object push model that is used in a limited way today.
- *Topology Management*: Enables the automatic configuration of the piconet topologies especially in scatternet situations that are becoming more common today. This should all be invisible to the users of the technology, while also making the technology just work.

- *QoS improvements*: Enable audio and video data to be transmitted at a higher quality, especially when best effort traffic is being transmitted in the same piconet.

7.10.2 UWB for AMP

The high speed (AMP) feature of Bluetooth 3.0 is based on 802.11, but the AMP mechanism was designed to be usable with other radios as well. It was originally intended for UWB, but the WiMedia Alliance, the body responsible for the flavor of UWB intended for Bluetooth, announced in March of 2009 that it was disbanding.

On March 16, 2009, the WiMedia Alliance announced it was entering into technology transfer agreements for the WiMedia Ultra-Wideband (UWB) specifications. WiMedia will transfer all current and future specifications, including work on future high speed and power optimized implementations, to the Bluetooth Special Interest Group (SIG), Wireless USB Promoter Group and the USB Implementers Forum. After the successful completion of the technology transfer, marketing and related administrative items

7.10.3 Communication and Connection

A master Bluetooth device can communicate with up to seven devices in a Wireless User Group. This network group of up to eight devices is called a piconet.

A piconet is an ad-hoc computer network, using Bluetooth technology protocols to allow one master device to interconnect with up to seven active devices. Up to 255 further devices can be inactive, or parked, which the master device can bring into active status at any time.

At any given time, data can be transferred between the master and one other device, however, the devices can switch roles and the slave can become the master at any time. The master switches rapidly from one device to another in a round-robin fashion. (Simultaneous transmission from the master to multiple other devices is possible, but not used much.)

The Bluetooth specification allows connecting two or more piconets together to form a scatternet, with some devices acting as a bridge by simultaneously playing the master role in one piconet and the slave role in another.

Many USB Bluetooth adapters are available, some of which also include an IrDA adapter. Older (pre-2003) Bluetooth adapters, however, have limited services, offering only the Bluetooth Enumerator and a less-powerful Bluetooth Radio incarnation. Such devices can link computers with Bluetooth, but they do not offer much in the way of services that modern adapters do.

7.10.4 Base Band Error Correction

Three types of error correction are implemented in Bluetooth systems,

- 1/3 rate (Forward Error Correction) (FEC)
- 2/3 rate FEC
- Automatic Repeat Request (ARQ)

Setting up Connections

Any Bluetooth device will transmit the following information on demand:

- Device name.
- Device class.
- List of services.
- Technical information, for example, device features, manufacturer, Bluetooth specification used, clock offset.

Any device may perform an inquiry to find other devices to connect to, and any device can be configured to respond to such inquiries. However, if the device trying to connect knows the address of the device, it always responds to direct connection requests and transmits the information shown in the list above if requested. Use of a device's services may require pairing or acceptance by its owner, but the connection itself can be initiated by any device and held until it goes out of range. Some devices can be connected to only one device at a time, and connecting to them prevents them from connecting to other devices and appearing in inquiries until they disconnect from the other device.

Every device has a unique 48-bit address. However these addresses are generally not shown in inquiries. Instead, friendly Bluetooth names are used, which can be set by the user. This name appears when another user scans for devices and in lists of paired devices.

Most phones have the Bluetooth name set to the manufacturer and model of the phone by default. Most phones and laptops show only the Bluetooth names and special programs are required to get additional information about remote devices. This can be confusing as, for example, there could be several phones in range named T610 (see Bluejacking).

7.10.5 Pairing

Pairs of devices may establish a trusted relationship by learning (by user input) a shared secret known as a **passkey**. A device that wants to communicate only with a trusted device can cryptographically authenticate the identity of the other device. Trusted devices may also encrypt the data that they exchange over the airwaves so that no one can listen in. The encryption can, however, be turned off, and passkeys are stored on the device file system, not on the Bluetooth chip itself. Since the Bluetooth address is permanent, a pairing is preserved, even if the Bluetooth name is changed. Pairs can be deleted at any time by either device. Devices generally require pairing or prompt the owner before they allow a remote device to use any or most of their services. Some devices, such as mobile phones, usually accept OBEX electronic business cards and notes without any pairing or prompts.

Certain printers and access points allow any device to use their services by default, much like unsecured Wi-Fi networks. Pairing algorithms are sometimes manufacturer-specific for transmitters and receivers used in applications such as music and entertainment.

Bluetooth 2.1 has an optional "touch-to-pair" feature based on Near Field Communication (NFC). By simply bringing two devices into very close range (around 10 cm/4 in), pairing can securely take place without entering a passkey or manual configuration.

7.10.6 Air Interface

The protocol operates in the license-free ISM band at 2.4-2.4835 GHz. To avoid interfering with other protocols that use the 2.45 GHz band, the Bluetooth protocol divides the band into 79 channels (each 1 MHz wide) and changes channels up to 1600 times per second. Implementations with versions 1.1 and 1.2 reach speeds of 723.1 kbit/s. Version 2.0 implementations feature Bluetooth Enhanced Data Rate (EDR) and reach 2.1 Mbit/s. Technically, version 2.0 devices have a higher power consumption, but the three times faster rate reduces the transmission times, effectively reducing power consumption to half that of 1.x devices (assuming equal traffic load).

7.10.7 Health Concerns

Bluetooth uses the microwave radio frequency spectrum in the 2.4 GHz to 2.4835 GHz range. Maximum power output from a Bluetooth radio is 100 mW, 2.5 mW, and 1 mW for Class 1, Class 2, and Class 3 devices respectively, which puts Class 1 at roughly the same level as mobile phones, and the other two classes much lower. Accordingly, Class 2 and Class 3 Bluetooth devices are considered less of a potential hazard than mobile phones, and Class 1 may be comparable to that of mobile phones.

7.10.8 Abbreviations and Acronyms

- *ACL*: Asynchronous Connectionless
- *ETSI*: European Telecommunications Standardization Institute
- *HCI*: Host Controller Interface
- *GAP*: Generic Access Profile
- *IrMC*: Infrared Mobile Communication Standard
- *ITU*: International Telecommunication Union
- *L2CAP*: Logical Link Control and Adaptation Layer
- *LMP*: Link Manager Protocol
- *ME*: Management Entity
- *OBEX*: Object Exchange Standard
- *PSM*: Protocol and Service Multiplexer
- *QoS*: Quality of Service
- *RFCOMM*: Radio Frequency Communication Protocol
- *SCO*: Synchronous Connection Oriented
- *SDL*: Specification and Description Language
- *SDP*: Service Discovery Protocol
- *TCS*: Telephony Control Protocol Specification
- *TS 07.10*: ETSI Technical Specification Standard
- *vCard, vCal*: IrDA based Applications
- *WAE*: Wireless Application Environment

- *WAP*: Wireless Application Protocol
- *WEP*: Wireless Encryption Protocol
- *WPAN*: Wireless Personal Area Network.

7.11 WIRELESS LOCAL LOOPS (WLL)

7.11.1 Introduction to WLL Technology

Mobile wireless has exploded in popularity because of the fact that it simplifies and revolutionizes communication. The market for mobile wireless is increasing by leaps and bounds. The success of mobile communications lies in the ability to provide instant connectivity anytime and anywhere and the ability to provide high-speed data services to the mobile user. The quality and speeds available in the mobile environment must match the fixed networks if the convergence of the mobile wireless and fixed communication networks is to happen in the real sense. So, the challenges for the mobile networks lie in providing a very large footprint of mobile services (to make the movement from one network to another as transparent to the user as possible) and the availability of high speed reliable data services along with high quality voice. A range of successful mobile technologies exists today in various parts of the world and every technology must evolve to fulfill all these requirements. In the following sections I'll talk about the mobile technologies existing today, how these technologies compare, how these technologies are shaping up and what we can expect to see in the near future.

WLL is often called the Radio Local Loop (RLL) or the Fixed Wireless Access (FWA). Since WLL is a kind of radio systems, its technology has been affected by wireless mobile communication technologies. The WLL systems have been developed according to the standards for second-generation cellular and cordless systems.

WLL systems tend to be used for one of the two reasons:

1. By monopoly or dominant public telephone network operators in regions with poor cable infrastructure as the most economic means of providing a basic telephone service.
2. By new competing operators in deregulated markets who are keen to build their own access network infrastructure rather than have to rely on the resources and services of their main competitor (the ex-monopoly carrier).

WLL systems for low speed telephony applications typically are designed to operate in the radio range 1-3 GHz. There are a number of point-to-multipoint systems which work in the 2.1 GHz, 2.2 GHz and 3.5 GHz bands.

The point-to-point systems can be classified into one of two broad types:

1. Systems including concentration functions, which connect multiple remote telephones by means of a radio link and a base station to a single V5.2 trunk.

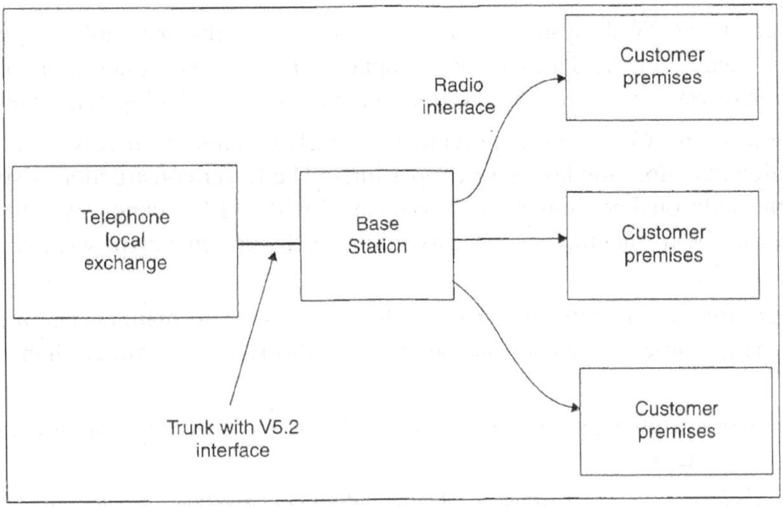

Fig. 7.9. With concentration-point to multi-point

2. Systems with no concentration function, Multiple-to-Point-Points (MPP)

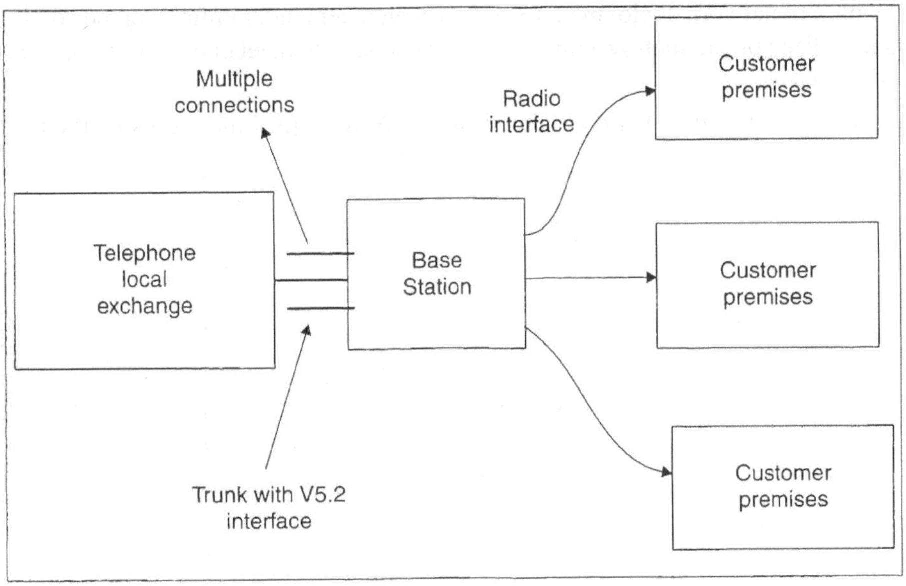

Fig. 7.10. Without concentration-multiple-point-to-point

The advantage of using a concentrating interface such as V5.2 is that fewer ports are required to connect the base station to the remote switch and less backhaul capacity is required on the trunk interconnecting the two. Wireless systems are justified as a local loop solution because of the cost-effectiveness and limitations of other technologies such as copper, coaxial cable and fiber.

7.11.2 Advantages of WLL

Fast Deployment: WLL systems can be deployed in weeks or months as compared to the months or years needed for the deployment of copper wire systems. Faster deployment can mean sooner realization of revenues and reduced time to payback of the deployment investment.

Low Construction Cost: The deployment of WLL technology involves considerably less heavy construction than does the laying of copper lines. The lower construction costs may be more than offset by the additional equipment costs associated with WLL technology, but in urban areas, especially, there may be considerable value in avoiding the disruption that the wide-scale deployment of copper lines entails.

Low Operation and Maintenance Cost: The operations and maintenance are easy and the average maintenance time per subscriber per year is shorter 3 to 4 times than their wire line competitors.

Customer Connection Cost: It is low, so overall cost per customer is significantly lower than wireline or cellular systems.

High Bandwidth Services Provision: Using advanced digital radio technologies, WLL can provide a variety of data services and multimedia services as well as voice.

Lower Network Extension Costs: Once the WLL infrastructure-network of base stations and interface to the telephone network-is in place, each incremental subscriber can be installed at very little cost. WLL systems that are designed to be modular and scalable can be furthermore allowing the pace of network deployment to closely match demand, minimizing the costs associated with the underutilized plant. Such systems are flexible enough to meet uncertain levels of penetration and rates of growth.

High System Capacity: Among radio systems, WLL enjoys the merits of fixed system.

UNIT 8 MOBILE DATA NETWORKS

8.1 INTRODUCTION

Wireless data communications was principally the domain of large companies with specialized needs; for example, large organizations that needed to stay in touch with their mobile sales force, or delivery services that needed to keep track of their vehicles and packages. However, this situation is steadily changing and wireless data communications is becoming as commonplace as its wired counterpart.

The need for wireless data communications arises partially because of the need for mobile computing and partially because of the need for specialized applications, such as computerized dispatch services and mobile fleet management.

Mobile computing, which aims to migrate the computing world onto a mobile environment, is affected primarily by two components: portability and connectivity. Portability, *i.e.,* the ability to untether computers from the conventional desktop environment, is getting increasingly feasible because with the continuous improvement in integration, miniaturization, and battery technology, the differences in performance and cost between desktop and portable computers are shrinking. Therefore, the processing power of desktop computing is becoming available to portable environments and this is highly desirable as far as productivity is concerned.

Regarding the connectivity, *i.e.,* the ability to connect to external resources and have access to external data, wireless data technology plays a significant part because it can offer ubiquitous connectivity, that is, connectivity at any place, any time. For this reason, wireless data technology can be of real value to the business world since computer users become more productive when they exploit the benefits of connectivity. The explosive growth of Local Area Network (LAN) installations over the past several years is ample evidence of the importance placed on connectivity by the business world.

Usually, portability and connectivity are at odds: the more portability increases, the more difficult it becomes to connect to external resources. However, wireless data technology provides the means to effectively combine both capabilities and, therefore, it is an essential technology for mobile computing.

Figure 8.1 presents the various wireless data technologies, which are essentially divided into two categories according to their mobility characteristics. For wide-area mobility there are mainly two available technologies: data transmission over cellular networks, whether analog or digital, and data transmission over mobile data networks. As shown in Fig. 8.1, the main difference between these two technologies is the data transport mode. Cellular networks, being primarily voice oriented, utilize circuit switching technology and, therefore, are optimized to isochronous data traffic conditions.

Whereas mobile data networks employ packet switching technology and are ideal for asynchronous data traffic transmission.

Currently, due to physical layer constraints, wide-area networks typically feature low-speed wireless data transmission, on the order of 9600 b/s. However, with the emerging new protocols, much higher data transmission speed is supported. For example, GPRS will support data transmission rates up to 115 kb/s, and HSCSD is designed to offer up to 56 kb/s over conventional voice channels.

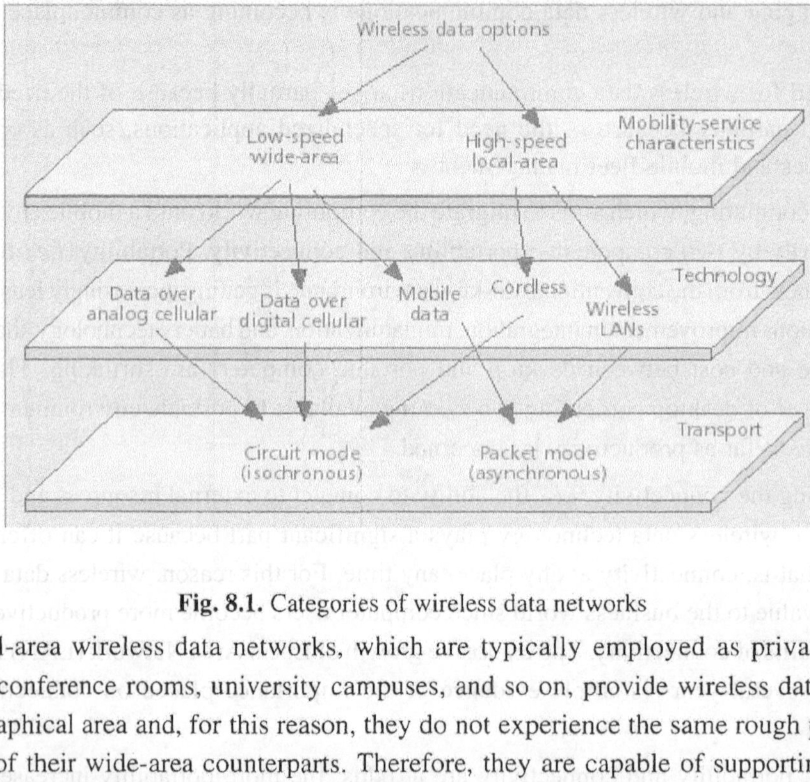

Fig. 8.1. Categories of wireless data networks

Local-area wireless data networks, which are typically employed as private systems in businesses, conference rooms, university campuses, and so on, provide wireless data service in a small geographical area and, for this reason, they do not experience the same rough physical layer constraints of their wide-area counterparts. Therefore, they are capable of supporting high-speed wireless data transmission, on the order of a few Mb/s. For local-area mobility there are mainly two alternatives: data transmission over cordless systems (*e.g.,* over CT-2 or DECT) and over wireless Local Area Networks (LANs). As indicated in Fig. 8.1, cordless systems provide circuit-switching transport service, whereas wireless LANs provides packet-switching transport service.

8.2 CDPD (CELLULAR DIGITAL PACKET DATA)

CDPD, cellular digital packet data, is a mobile data technology that permits subordinate data operation on the spectrum assigned to the Advanced Mobile Phone Service (AMPS). It was first introduced by IBM as a packet-switching overlay to the existing analog cellular voice network and frequencies. Later, a CDPD system specification was created by a consortium of cellular carriers including AirTouch, McCaw Cellular, Southwestern Bell Mobile Systems, NYNEX, Ameritech, GTE, Bell Atlantic Mobile, and Contel Cellular. CDPD technology is being deployed by a number of cellular companies in the U.S., including Bell Atlantic, Ameritech, GTE, and McCaw Cellular. Equipment is provided by a variety of manufacturers.

CDPD systems are designed to take advantage of the idle voice channels of an analog cellular network, such as AMPS. These idle channels are used to transmit short data messages and establish a packet-switching service. In order to utilize those idle channels, CDPD implements a hopping procedure among the available cellular frequencies. The air interface operates at a raw data rate of 19.2 kb/s and provides forward error correction to combat the interference and fading of the cellular channels.

The Wireless Data Forum is an industry association that handles the shaping of the CDPD technology and supports the growth of the commercial marketplace. This forum aims to help the operators, equipment providers, and billing system developers through the complicated issues they face. Among other things, the forum tries to help all of the CDPD operators to develop inter-operator roaming and invoicing.

According to the Wireless Data Forum, by the end of the third quarter 1998, CDPD was available in 195 markets in the United States: 118 Metropolitan Statistical Areas (MSAs), 41 Rural Statistical Areas (RSAs), and 36 international markets, and was available to 53 percent of the U.S. population.

8.2.1 System Description of CDPD

The primary elements of a CDPD network are the End Systems (ESs) and the Intermediate Systems (ISs), as shown in Fig. 8.2. The ESs represent the actual physical and logical end nodes that exchange information, while the ISs represent the CDPD infrastructure that stores, forwards, and routes the information.

There are two kinds of ESs: The Mobile End System (M-ES), which is a device used by a subscriber to access the CDPD network over the wireless interface, and the Fixed End System (F-ES), which is a common host, server, or gateway that supports or provides access to data and applications. By definition, the location of an F-ES is fixed, whereas the location of an M-ES may change.

Typically, each M-ES consists of a mobile terminal (personal computer, personal digital assistant, or other standard device), and an additional device, the radio modem, that attaches to the mobile terminal and manages the radio links and protocols. These devices usually communicate over standard serial protocols, such as the Serial Line Internet Protocol (SLIP), or the Point-to-Point (PPP) protocol.

On the other hand, there are two kinds of ISs: a "generic" IS, which is simply a (IP) router that has no knowledge of CDPD and mobility issues, and a Mobile Data Intermediate System (MD-IS), which is a specialized IS that routes messages based on its knowledge of the current location of an M-ES. More specifically, a MD-IS is a set of hardware components and software functions that provide switching, accounting, registration, authentication, encryption, and mobility management functions. The mobility management software allows the switching system to track M-ESs regardless of their location in the network, and allows M-ESs to use a single network address. The CDPD mobility management software follows the mobile-IP model, established by the Internet Engineering Task Force (IETF). Besides the ESs and the ISs, there is also another element called the Mobile Data Base Station (MDBS), which is analogous to the AMPS base station. A MDBS is a combination of a computer, power amplifiers, and a radio transceiver. It performs no networking functions but it is a link-layer relay; it sends and receives information from the M-ESs and relays it back to the MD-IS. It also controls the radio interface and manages the radio communications, and monitors the activity on the voice network (to ensure that data and voice do not interfere with each other). The MDBS creates an air link comprising two RF channels for forward and reverse communications with multiple M-ESs.

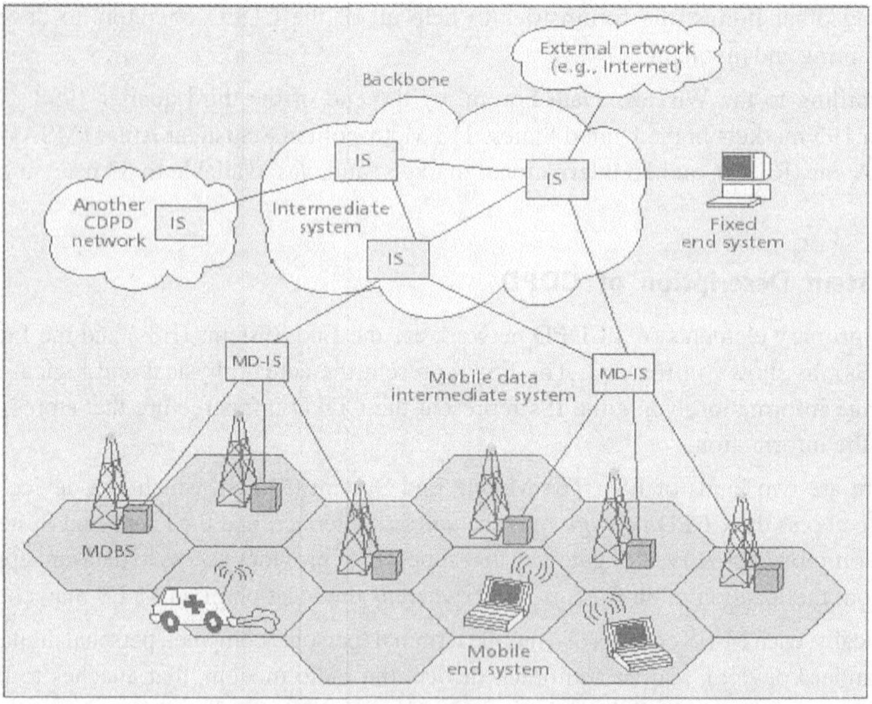

Fig. 8.2. CDPD network architecture

Fig. 8.3 shows the standardized interfaces used across the CDPD network. The M-ESs are connected to the CDPD network through the A-interface (the air interface), while the F-ESs are connected through the E-interface. The E-interface is also used to interconnect with external networks. Finally, the I-interface is used in the backbone, between the various ISs, and at the interconnection points with other CDPD networks. All these interfaces and the associated protocols are extensively described in.

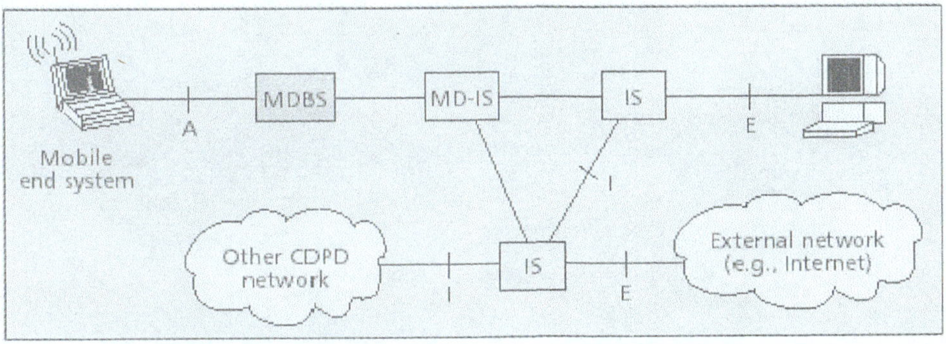

Fig. 8.3. Standardized interfaces within CDPD architecture

The CDPD backbone provides connectionless transport services, also called "datagram" services. This means that the network individually routes packets, based on the destination address the packet carries and on the knowledge of the current network topology. For the routing of packets, CDPD supports both the Internet Protocol (IP) and the Connectionless Network Protocol (CLNP), which is an OSI standard protocol.

The CDPD system is designed to facilitate the interoperability between the networks of different service providers . Especially in CDPD, where there may be many different operators in a given geographical area (since there may also be many different cellular operators, each one deploying its own CDPD infrastructure), interoperability is a main concern. Interoperability means that all subscriber equipment, components, functions, and processes of the CDPD system work together within all the cooperating CDPD networks. Interoperability allows for seamless roaming inside a large geographical area and provides the means for ubiquitous CDPD service. In fact, the CDPD network is envisioned as an internet work composed of multiple administrative domains, each one operated by a different service provider (see Fig. 8.4).

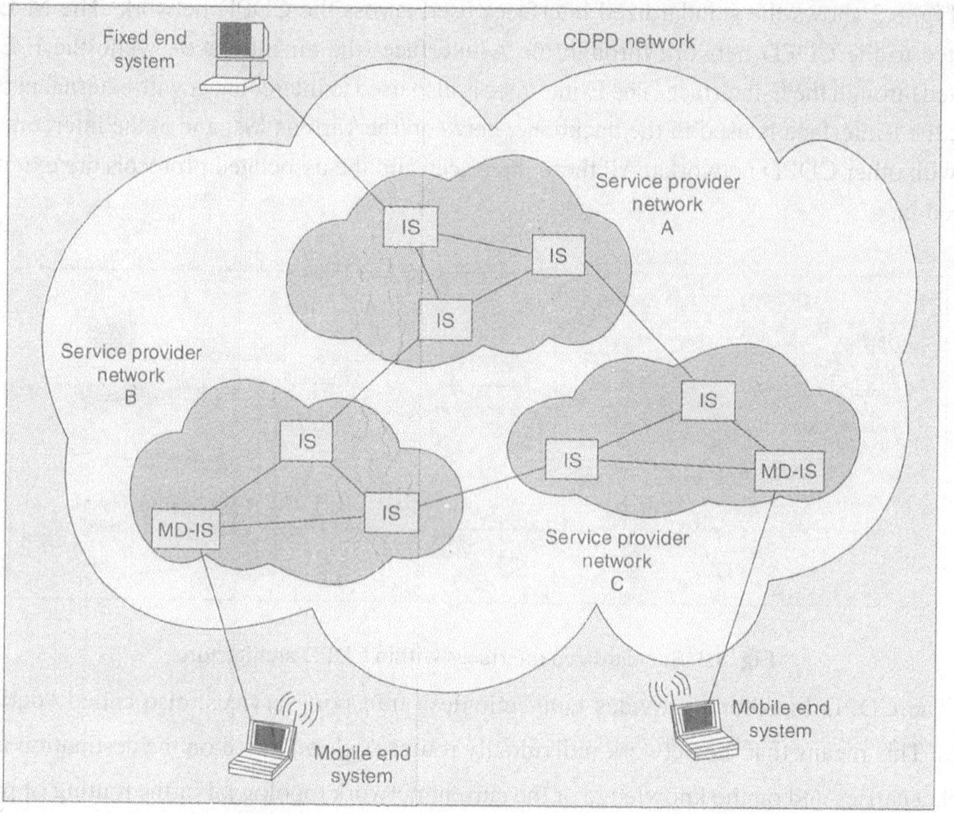

Fig. 8.4. CDPD service with interoperability between service providers

8.2.2 Protocol Architecture of the CDPD

The protocol architecture of the CDPD air interface is illustrated in Fig. 8.5. The physical (PHY) layer in CDPD corresponds to a functional entity that accepts a sequence of bits from the Medium Access Control (MAC) layer and transforms them into a modulated waveform for transmission onto a physical 30 kHz RF channel. As illustrated in Fig. 8.5, communication between a MDBS and a M-ES takes place over a pair of such RF channels (having a fixed frequency separation). The first channel, called the forward channel, accommodates transmissions in the direction from the MDBS to the M-ESs and is either dedicated to CDPD use or shared with the voice cellular network. In any case, transmission on the forward channel is continuous as long as it is in use for CDPD. The second channel, called the reverse channel, accommodates transmissions in the direction from the M-ESs to the MDBS and is shared among all M-ESs communicating with the same MDBS. A pair of associated reverse and forward channels forms a CDPD channel stream.

The PHY interfaces with another entity, the Radio Resource Management Entity (RRME) through this control interface the RRME can:

- Tune the physical layer to a specific RF channel pair.
- Set the transmission power level to the desired value.

- Measure the received signal level of a RF channel and estimate its potential to offer acceptable communication.
- Suspend and resume the operation of the physical layer in cases where powers saving facilities are required.

The modulation employed on a RF channel stream is GMSK with BT = 0.5. A frequency greater than the central carrier frequency represents a logical 1, while a logical 0 is represented by a frequency less than the central carrier frequency. The modulation rate on both the forward and reverse RF channels is 19.2 kb/s.

The MAC layer models a functional entity logically operating between the PHY and LLC layers. The MAC layer within an M-ES cooperates with the corresponding MAC layer within the MDBS. The purpose of this layer is to convey information, namely, link protocol data units (LPDUs), between peer LLC entities across the CPDP air interface. For this purpose, the MAC layer provides the following services:

- Encapsulates LPDUs into frame structures to ensure LPDU delimiting, frame synchronization, and data transparency.
- Encodes LPDUs to provide error protection against mobile channel impairments.
- Detects and corrects bit errors within received frames.
- Arbitrates access to the shared reverse channel.
- Synchronizes with the forward channel transmissions to make feasible the reception of data as well as control information transmitted in every CDPD cell.

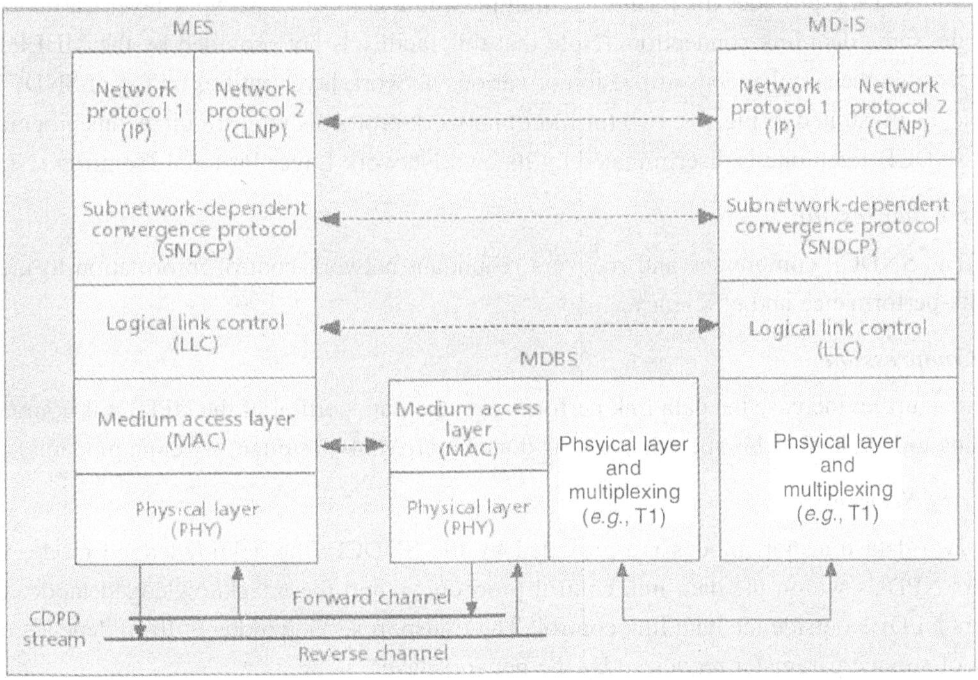

Fig. 8.5. Protocol architecture of CDPD air interface

The Subnetwork Dependent Convergence Protocol (SNDCP) lies between the data link layer and the network layer. The latter is assumed to be subnetwork independent, *i.e.,* it is built to work over virtually any data link and therefore it does not take into account the specific features of the MDLP. For this reason, the services assumed by the network protocol(s) may not map directly into the services provided by MDLP. In this case, the SNDCP is operated to provide the required cooperation.

More specifically, the SNDCP provides the following functions:

Segmentation

Network Protocol Data Units (NPDUs) are segmented and reassembled where needed in order to be accommodated within the limited length of the data link frames. With this segmentation, the maximum size of a NPDU can be 2048 bytes, while the maximum size of user data supported by MDLP is considerably smaller (default value is 130bytes).

Encryption

To provide user data confidentiality over the CDPD air interface the NPDUs are encrypted after being segmented. The secret keys used for the encryption and decryption are obtained by means of a Security Management Entity (SME) that operates on top of SNDCP as a network layer entity.

Multiplexing

The SNDCP provides the means for multiplexing a number of network-layer traffic streams within the same data link connection. (Note that this facility is not provided by the MDLP.) This makes feasible the simultaneous utilization of various network-layer entities on top of SNDCP. For example, as illustrated in Fig. 10, two (or more) network protocols may simultaneously operate on top of SNDCP. Each one is discriminated by its own Network Layer Protocol Identifier (NLPI).

Header Compression

The SNDCP compresses and recovers redundant network-control information to increase data link performance and efficiency.

Data Compression

To further increase the data link performance, the data portion of the NPDUs is compressed according to ITU-T V.42 bis specification (as done by all V.34-compliant wireline modems).

Quality of Service

Two data transport modes are provided by the SNDCP: the acknowledged mode, which transfers NPDUs within the data link control procedures, and the unacknowledged mode, which transfers NPDUs outside the data link control. The transport service mode utilized depends on the quality of service parameter requested by the network layer.

8.3 GPRS (GENERAL PACKET RADIO SERVICE)

Its introduction is one of the key steps in the evolution of today's GSM networks to the third generation, and GSM operators around the world are upgrading their networks with a view to launching commercial GPRS services in 2000. The main feature of GPRS is that it reserves radio resources only when there is such a need and that these radio resources are shared by all mobile stations (MSs) in a cell. Therefore, as is the case for all packet data services, effective resource utilization is provided for bursty data applications, such as telemetry, train control systems, interactive data access, toll road charging systems, and Internet browsing using the World Wide Web. Data rates of up to 115kb/swouldbesupported.

The main objective of GPRS is to offer a mobile packet interface to standard data networks (such as TCP/IP, X.25, and CLNP). This interface should be embedded in the conventional GSM network architecture, which was originally designed for circuit-switched integrated services. For this purpose, GPRS introduces some new functional elements to the general GSM architecture and it modifies some mobility management functions. Some cooperation still exists between the elements of the standard GSM services and GPRS. In comparison with CDPD, it is important to note that GPRS provides a data overlay within the standard GSM infrastructure (with some new functional elements), while CDPD provides a packet data overlay but with a totally separate infrastructure.

8.3.1 System Description

The GPRS reference model is depicted in Fig. 8.6. GPRS allows the subscriber to send and receive data in an end-to-end packet transfer mode, without using any network resources in circuit-switched mode. This allows for autonomous operation of GPRS and best fits the bursty traffic characteristics. Packet routing and transfer within the public land mobile network (PLMN) is supported by definition of a new logical network node called a GPRS support node (GSN). The GSN is basically a packet router with additional mobility management features and connects with various network elements through standardized interfaces. The GSN node that acts as a physical interface to the external packet data networks (PDNs) is called as the gateway GSN (GGSN), whereas the GSN node that connects with a BSC (base station controller) and directly handles packet delivery to and from MSs is called as the serving GSN (SGSN). Each SGSN is responsible for the delivery of packets to the MSs within its service area.

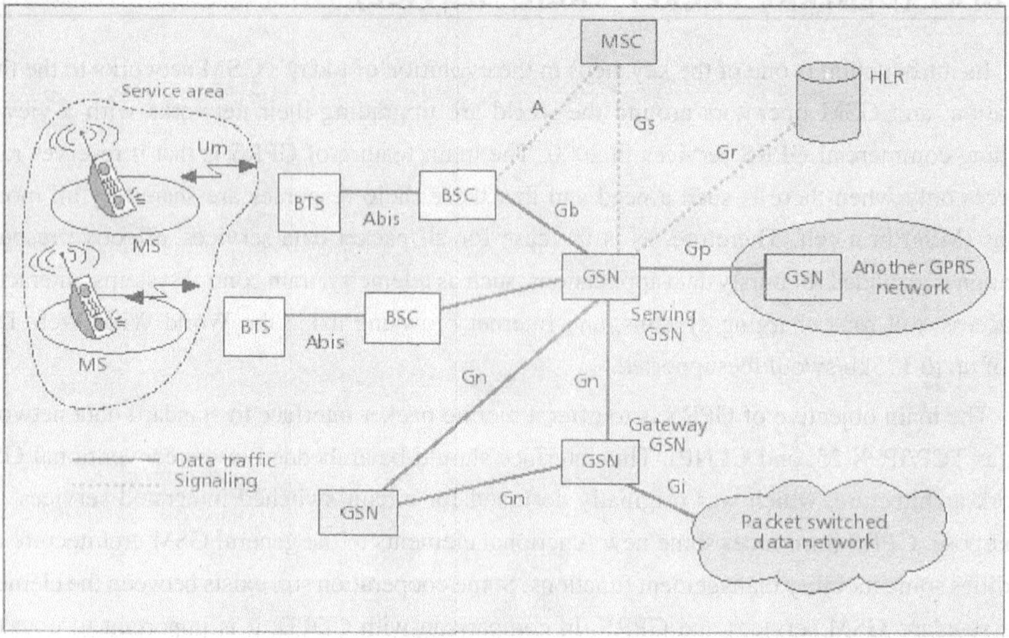

Fig. 8.6. GPRS system architecture

The general packet routing procedure is performed as illustrated in Fig. 8.7. In the case where a mobile originates traffic for a fixed host, the SGSN encapsulates the data packets and routes them to the appropriate GGSN, where they are forwarded to the correct packet switched data network (PSDN). Specific routing policies are applied inside this PSDN to send the packets to the corresponding host. On the other hand, packets coming from a corresponding host are first routed to the GGSN through the PSDN, based on the examination of the destination address. The GGSN checks the routing context associated with this destination address and determines the address of the SGSN currently serving the addressed MS. Subsequently, the original data packet is encapsulated into another packet (this procedure is called tunneling), which is forwarded to the SGSN and ultimately delivered to the correct MS. The GPRS backbone network is a private IP network. The IP addresses used in this backbone are selected by the GPRS operator and they are not known outside the PLMN. Routing between GPRS mobile terminals is usually accommodated through inter-PLMN packet networks, as illustrated in Fig. 8.7.

Within GPRS, two different encapsulation schemes are used. First, between the GSNs all packets are encapsulated by means of a GPRS tunneling protocol (GTP) in order to enable usage of different packet data protocols, even if these protocols are not supported by all SGNSs. Second, encapsulation between the MS and SGSN is performed to decouple the logical link management from the network-layer protocols.

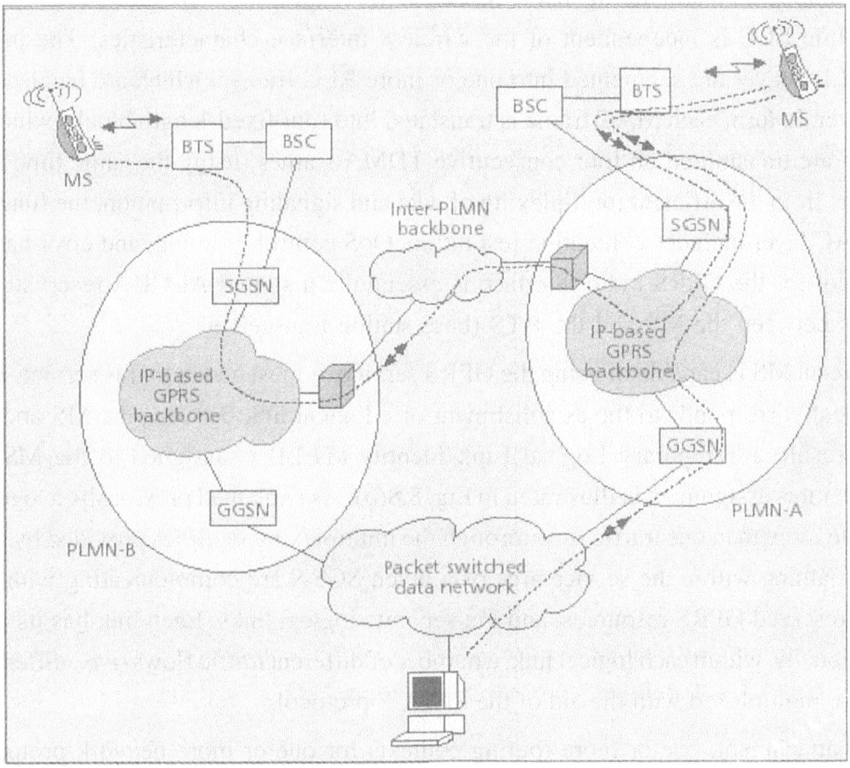

Fig. 8.7. Typical routing within GPRS

8.3.2 Protocol Architecture

The GPRS service is based on the layered architecture shown in Fig. 8.8(*a*). At the top of the protocol layer, there may be a number of network protocols, which are treated as prospective users of the GPRS service. The Sub Network-Dependent Convergence Protocol (SNDCP) provides a multiplexing layer that aids the transmission of multiple network layer messages onto a single logical link connection. Furthermore, the SNDCP includes ciphering, segmentation, and compression facilities. Its functionality is very similar to the functionality of the SNDCP layer in CDPD, explained previously.

The Logical Link Control (LLC) layer, at the bottom of the SNDCP layer, provides a logical link between the MS and SGSN. Protocol functionality is based on LAPD as used within the GSM signaling plane, but additional features are supported, e.g., point-to-multipoint transmission. This link control protocol is sometimes called Link Access Procedure on the G-channel (LAPG).

The Radio Link Control/Medium Access Control (RLC/MAC) layer arbitrates access to the shared medium between the MSs and the network, and also provides a reliable link between an MS and a base station. It is worth noting that RLC is specific to the radio technology used in the wireless interface, while LLC is independent of the wireless interface characteristics. The protocol data units at the LLC layer are segmented into one or more RLC frames, which are handed over to the MAC sublayer. In turn, each MAC frame is translated into four fixed-length blocks which, after bit-interleaving, are transmitted on four consecutive TDMA frames (using the same time slot in each frame). Apart from the efficient multiplexing of data and signaling information, the functionality of the RLC/MAC layer includes contention resolution, QoS control, framing, and error handling. The MAC protocol for the GPRS radio interface is essentially a slotted ALOHA reservation protocol and operates between the MS and the BTS (base station transceiver).

Before an MS is capable of using the GPRS service, it must attach to this service. Effectively, this attachment corresponds to the establishment of a logical link between the MS and its serving GSN. As a result, a Temporary Logical Link Identity (TLLI) is assigned to the MS. A logical condition after this assignment is illustrated in Fig. 8.8(*b*). As explained previously, a logical link can accommodate more than one traffic flow through the multiplexing facilities provided by the SNDCP layer. MSs roaming within the service area of a given SGSN are communicating with this SGSN through the reserved GPRS resources, and via separate logical links. Each link has its own unique TLLI. Additionally, within each logical link, a number of different traffic flows (*e.g.*, different network protocols) are multiplexed with the aid of the SNDCP protocol.

After attachment, one or more routing contexts for one or more network protocols can be negotiated with the SGSN. In order to verify that a given MS is allowed to use a network protocol, the Home Location Register (HLR) is queried. Among other things, the subscription profile found in the HLR includes the matching GGSN address. If access is permitted, the GGSN is requested to update the routing context (*i.e.,* the serving SGSN address and tunneling information) accordingly. As explained before, this information is used to support the routing procedure in a mobile environment.

During the GPRS session, the location of an MS is being tracked. When in ready state (*i.e.,* during active communication), the MS informs the SGSN about every cell change. However, when in standby state (*i.e.,* when waiting for an event to trigger transmission/reception), the MS requests location updates upon a Routing Area (RA) change. The routing area consists of an operator-defined group of cells. If a RA update takes place and the new RA is handled by another SGSN, then the new SGSN asks the old SGSN to send the mobility management information of the MS. Subsequently, the GGSN and the HLR are updated with the new routing context, and the old SGSN deletes the corresponding information of this MS.

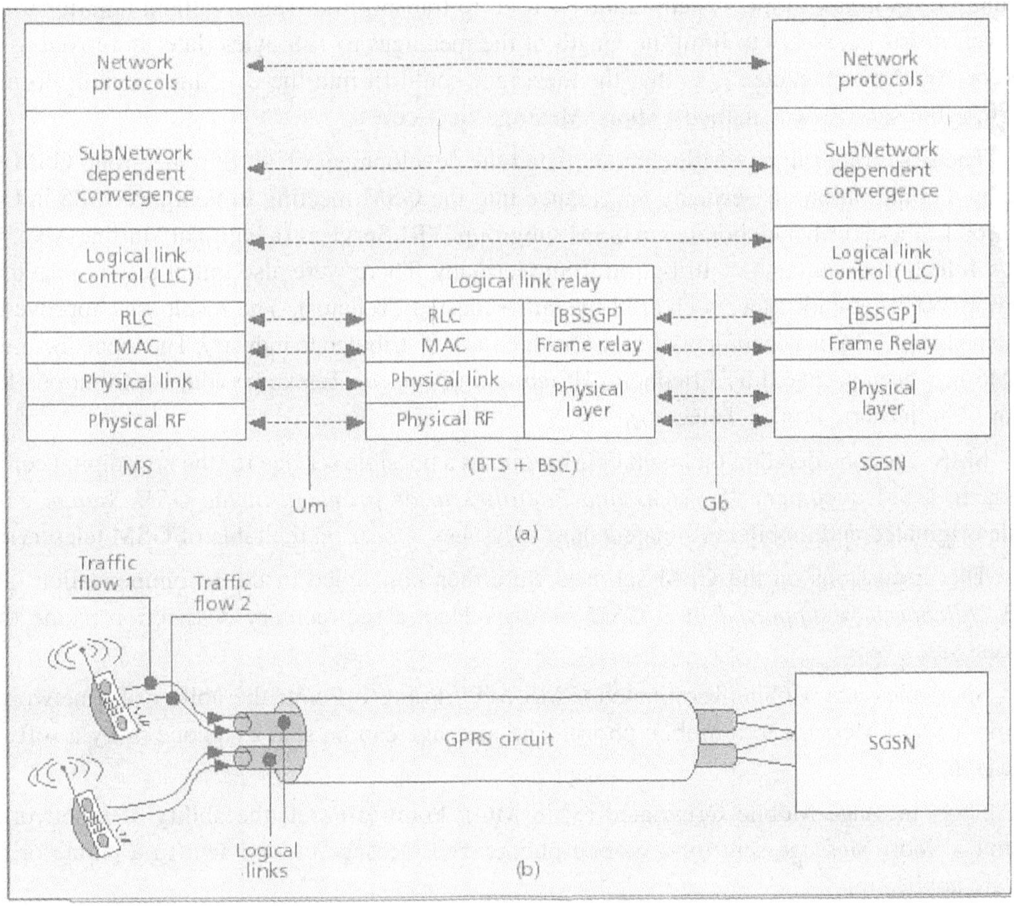

Fig. 8.8. (*a*) GPRS protocol architecture;
(*b*) schematic presentation of traffic and logical link multiplexing in GPRS

8.4 SMS AS PART OF GSM

The idea of adding text messaging to the services of mobile users was latent in many communities of mobile communication services at the beginning of the 1980s. The first action plan of the CEPT Group GSM approved in December 1982 requested "The services and facilities offered in the public switched telephone networks and public data networks ...should be available in the mobile system". This target includes the exchange of text messages either directly between mobile stations or the transmission via Message Handling Systems widely in use since the beginning of the 1980s.

The innovation in SMS is indicated by the word **Short** in Short Message Service. The GSM system is optimized for telephony, since this was identified as its main application. The key idea for SMS was to use this telephony-optimized system and to transport messages on the signaling paths needed to control the telephony traffic during time periods when no signaling traffic existed. In this way unused resources in the system could be used to transport messages without additional cost. However, it was necessary to limit the length of the messages to 128 bytes (later improved to 140 bytes, or 160 7-bit characters), so that the messages could fit into the existing signaling formats. Therefore the service was named "**Short** Message Service".

The first proposal for SMS which initiated the development of SMS in the group GSM was made by a contribution of Germany and France into the GSM meeting in February 1985 in Oslo. This proposal was further elaborated in GSM subgroup WP1 Services (Chairman Martine Alvernhe, France Telecom) based on a contribution from Germany. There were also initial discussions in the subgroup WP3 network aspects chaired by Jan Audestad (Telenor). The result was approved by the main group GSM in a document of June 85 which was distributed to industry. The input documents on SMS had been prepared by Friedhelm Hillebrand (Deutsche Telekom) with contributions from Bernard Ghillebaert (France Télécom).

SMS was considered in the main GSM group as a possible service for the new digital cellular system. In GSM document "*Services and Facilities to be provided in the GSM System*", both mobile originated and mobile terminated short messages appear on the table of GSM teleservices.

The discussions on the GSM services were then concluded in the recommendation GSM 02.03 "*TeleServices supported by a GSM PLMN*". Here a rudimentary description of the three services was given:

Short message Mobile Terminated (SMS-MT)/ Point-to-Point: the ability of a network to transmit a Short Message to a mobile phone. The message can be sent by phone or by a software application.

Short message Mobile Originated (SMS-MO)/ Point-to-Point: the ability of a network to transmit a Short Message sent by a mobile phone. The message can be sent to a phone or to a software application.

8.4.1 Short Message Service (SMS)

The *Short Message Service-Point to Point (SMS-PP)* is defined in GSM recommendation 03.40.GSM 03.41 defines the *Short Message Service-Cell Broadcast* (*SMS-CB*) which allows messages (advertising, public information, etc.) to be broadcast to all mobile users in a specified geographical area. Messages are sent to a Short Message Service Centre (SMSC) which provides a store-and-forward mechanism. It attempts to send messages to their recipients. If a recipient is not reachable, the SMSC queues the message for later retry. Some SMSCs also provide a "forward and forget" option where transmission is tried only once. Both Mobile Terminated (MT), for messages sent to a mobile handset, and Mobile Originating (MO), for those that are sent from the mobile handset, operations are supported. Message delivery is best effort, so there are no guarantees that a message will actually be delivered to its recipient and delay or complete loss of a message is not uncommon, particularly when sending between networks. Users may request delivery reports to confirm that messages reach the intended recipients, either via the SMS settings of most modern phones, or by prefixing each message with *0# or *N#.

Message Size

Transmission of short messages between the SMSC and the handset is done using the Mobile Application Part (MAP) of the SS7 protocol. Messages are sent with the MAP mo- and mt-ForwardSM operations, whose payload length is limited by the constraints of the signaling protocol to precisely 140 octets (140 octets = 140 * 8 bits = 1120 bits). Short messages can be encoded using a variety of alphabets: the default GSM 7-bit alphabet (shown above), the 8-bit data alphabet, and the 16-bit UTF-16/UCS-2 alphabet. Depending on which alphabet the subscriber has configured in the handset, this leads to the maximum individual Short Message sizes of 160 7-bit characters, 140 8-bit characters, or 70 16-bit characters (including spaces). Support of the GSM 7-bit alphabet is mandatory for GSM handsets and network elements, but characters in languages such as Arabic, Chinese, Korean, Japanese or Cyrillic alphabet languages (*e.g.* Russian) must be encoded using the 16-bit UCS-2 character encoding (see Unicode). Routing data and other metadata is additional to the payload size.

Larger content (Concatenated SMS, multipart or segmented SMS or "long sms") can be sent using multiple messages, in which case each message will start with a user data header (UDH) containing segmentation information. Since UDH is inside the payload, the number of characters per segment is lower: 153 for 7-bit encoding, 134 for 8-bit encoding and 67 for 16-bit encoding. The receiving handset is then responsible for reassembling the message and presenting it to the user as one long message. While the standard theoretically permits up to 255 segments, 6 to 8 segment messages are the practical maximum, and long messages are often billed as equivalent to multiple SMS messages. See Concatenated SMS for more information. Some providers have offered length-oriented pricing schemes for SMSs; however, the phenomenon is disappearing.

AT Commands

Many mobile and satellite transceiver units support the sending and receiving of SMS using an extended version of the Hayes command set. The connection between the Terminal Equipment and the transceiver can be realized with a serial cable (*i.e.* USB), a Bluetooth link, an infrared link, etc. Common AT commands include AT+CMGS (send message), AT+CMSS (send message from storage), AT+CMGL (list messages) and AT+CMGR (read message).

However, not all modern devices support receiving of messages if the message storage, for instance the device's internal memory, is not accessible using AT commands.

8.5 MOBILE APPLICATION PROTOCOL

The Mobile Application Part (MAP) is an SS7 protocol which provides an application layer for the various nodes in GSM and UMTS mobile core networks and GPRS core networks to communicate with each other in order to provide services to mobile phone users. The Mobile Application Part is the application-layer protocol used to access the Home Location Register, Visitor Location Register, Mobile Switching Center, Equipment Identity Register, Authentication Centre, Short message service center and Serving GPRS Support Node.

Facilities Provided

The primary facilities provided by MAP are:

- Mobility Services: location management (roaming), authentication, managing service subscription information, fault recovery,
- Operation and Maintenance: subscriber tracing, retrieving a subscriber's IMSI
- Call Handling: routing, managing calls whilst roaming, checking that a subscriber is available to receive calls
- Supplementary Services
- Short Message Service
- Packet Data Protocol (PDP) services for GPRS: providing routing information for GPRS connections
- Location Service Management Services: obtaining the location of subscribers

MAP Signaling

In mobile cellular telephony networks like GSM and UMTS the SS7 application MAP is used. Voice connections are Circuit Switched (CS) and data connections are Packet Switched (PS) applications.

Some of the **GSM/UMTS Core Switched interfaces** in the Mobile Switching Center (MSC) transported over SS7 include the following:

B -> VLR (uses MAP/B). Most MSCs are associated with a Visitor Location Register (VLR), making the B interface "internal".

D -> HLR (uses MAP/D) for attaching to the CS network and location update.

E -> MSC (uses MAP/E) for inter-MSC handover.

F -> EIR (uses MAP/F) for equipment identity check.

H -> SMS-G (uses MAP/H) for Short Message Service (SMS) over CS.

There are also several **GSM/UMTS PS interfaces** in the Serving GPRS Support Node (SGSN) transported over SS7:

Gr -> HLR for attaching to the PS network and location update.

Gd -> SMS-C for SMS over PS.

Gs -> MSC for combined CS+PS signaling over PS.

Ge -> Charging for Customized Applications for Mobile networks Enhanced Logic (CAMEL) prepaid charging.

Gf -> EIR for equipment identity check.

Index